MANIFOLDS II - THEORY AND APPLICATIONS

Edited by **Paul Bracken**

Manifolds II - Theory and Applications
http://dx.doi.org/10.5772/intechopen.75217
Edited by Paul Bracken

Contributors

Weifeng Liu, Xueqi Ma, Yuichi Kobayashi, Mehmet Akyol, Samuel Kadoury, Zhengsheng Chen, Paul Bracken

First published in London, United Kingdom, 2019 by IntechOpen
IntechOpen is the global imprint of INTECHOPEN LIMITED, registered in England and Wales, registration number: 11086078, The Shard, 25th floor, 32 London Bridge Street
London, SE19SG – United Kingdom
Printed in Croatia

British Library Cataloguing-in-Publication Data
A catalogue record for this book is available from the British Library

Additional hard copies can be obtained from orders@intechopen.com

Manifolds II - Theory and Applications, Edited by Paul Bracken
p. cm.
Print ISBN 978-1-83880-309-4
Online ISBN 978-1-83880-310-0
eBook (PDF) ISBN 978-1-83880-716-0

For EU product safety concerns:
IN TECH d.o.o., Prolaz Marije Krucifikse Kozulić 3, 51000 Rijeka, Croatia,
info@intechopen.com or visit our website at intechopen.com.

Meet the editor

Paul Bracken is a professor in the Department of Mathematics at the University of Texas in Edinburg, TX. His BSc degree is from the University of Toronto and he has a PhD degree from the University of Waterloo, Canada. His research interests are quite broad and include studying problems in mathematical physics such as quantum mechanics and quantum field theory. He has also worked in the areas of partial differential equations and differential geometry, as well as gravity. He has written some 160 chapters and several more that have appeared in books. He has also published two short books and edited several books for IntechOpen publisher, including this one, which is the fourth volume that he has contributed to with IntechOpen. He is on the editorial boards of several journals, has presented many talks at various meetings and conferences, and has taught courses in mathematics and physics at all levels over the years.

Contents

Preface

This volume is a collection of chapters dedicated to the investigation of manifolds. Research in this area, in particular differentiable manifolds, which often have physical applications, forms an integral part of mathematics research. In addition to the significant interest manifolds hold for pure mathematicians, manifolds also have very important applications to many areas of modern applied mathematics and the physical sciences. As the book will show, there are numerous applications to such diverse areas as partial differential equations, dynamical systems, and even constructing computer images. *Manifolds II: Theory and Applications* is basically divided into two groups. The first part is broken down into a group of three chapters underlying theoretical aspects of manifolds and a group of three chapters directed toward applications of manifolds to applied areas of science.

The first group presents chapters of a theoretical nature on the ideas behind manifold regularization and conformal anti-invariant submersions whose total manifolds are locally product Riemannian. There is also a chapter on the generalized Weierstrass system for inducing mean curvature surfaces in Euclidean three-space. This area has seen a lot of activity recently and the chapter is written with both mathematicians and physicists in mind.

The last three chapters form a collection of chapters that touch on manifolds in a very applied manner, such as manifold-based robot motion generation.

The book has been put together by an international group of invited authors. It is a pleasure to thank them for their hard work and significant contributions. I gratefully acknowledge the assistance provided by Mr. Nino Popović, who was the author service manager throughout the publishing process, as well as the IntechOpen for the opportunity to edit this volume that examines the subject of manifolds.

Professor Paul Bracken
Department of Mathematics
University of Texas, USA

Section 1

Theoretical

Chapter 1

The Generalized Weierstrass System in Three-Dimensional Euclidean Space

Paul Bracken

Additional information is available at the end of the chapter

http://dx.doi.org/10.5772/intechopen.82631

Abstract

In this chapter, some recent advances in the area of generalized Weierstrass representations will be given. This is an approach to the theory of surfaces in Euclidean three space. Weierstrass representations permit the explicit construction of surfaces in the designated space. The discussion proceeds in a novel and introductory manner. The inducing formulas for the coordinates of a surface are derived and important conservation laws are formulated. These lead to the inducing mechanism of a surface in terms of solutions to a system of two-dimensional Dirac equations. A set of fundamental forms as well as expressions for the mean and Gaussian curvatures are derived. The Cartan moving frame picture is also formulated to put everything in a broader perspective. A connection with the nonlinear sigma model is presented, which has important applications in physics. Some relationships are established between integrable systems and geometry by way of conclusion.

Keywords: metric, tensor, manifold, Weierstrass representation, curvature, evolution equation
Mathematics Subject Classification: 35Q51,53A10

1. Introduction

The theory of immersions and deformations of surfaces has been an important area of study as far as classical differential geometry is concerned. An inducing mechanism for describing minimal surfaces imbedded in three-dimensional Euclidean space was first put forward by Enneper and Weierstrass in the nineteenth century [1]. Their basic ideas have been extended and generalized by Konopelchenko and colleagues [2–4]. The connection between certain classes of constant mean curvature surfaces and the trajectories of an infinite-dimensional Hamiltonian system was put forward first by Konopelchenko and Taimanov [2], and has proved to be very useful in investigating types of questions related to this and other types of spaces and in higher dimensions [5, 6].

Surfaces and their dynamics play a very crucial and important role in a great number of phenomena which arise in the physical sciences in general. A longer introduction and more examples can be found in [7, 8]. They appear in the study of surface waves, shock waves, deformations of membranes, as well as in many problems in hydrodynamics connected with the motion of boundaries between regions of differing densities and viscosities. At the present time, they are appearing in string theory models [9–11] and in the study of integrable systems in general [12, 13]. A special case is that of surfaces which have zero mean curvature. These surfaces are usually referred to as minimal surfaces. The work of Weierstrass and Enneper originally concerned itself with the construction of minimal surfaces in three-dimensional Euclidean space [14, 15].

It is the intention here to present an introduction to the work of Konopelchenko and referred to presently as the generalized Weierstrass representation. The work presents both mathematical and physical developments in the area which should be relevant to both physicists and mathematicians. The development starts by studying a coupled system of two-dimensional Dirac equations in terms of two complex functions that involves a mass term that depends on two coordinates of the space. This equation can then be decomposed into a system of two simpler equations and their respective complex conjugates. By looking at such things as conservation laws, inducing formulas which specify the coordinates of a surface in Euclidean three space can be deduced, as well as the first and second fundamental forms pertaining to the surface. A remarkable result of this development is that the mass which appears in the Dirac system becomes related to the mean curvature of the surface. One might say this indicates that mass is a consequence of geometry in this type of model. To fit these developments in the larger picture of modern differential geometry, the Cartan moving frame for the system is formulated out of which emerges another remarkable result. Namely, the two-dimensional Dirac equation is a way of writing an affine connection on the surface. Finally, by investigating the Gauss map, it is shown that there is a mathematical way of proceeding from the Dirac system and the nonlinear sigma model in two dimensions [16, 17]. The whole construction leads to a very deep link between nonlinear evolution equations and geometry as a whole [18, 19]. The paper finishes with some interesting examples and outlook for further work.

2. Two-dimensional Dirac equation and construction of surfaces

The process of inducing surfaces in three-dimensional space can be generalized by establishing a system of Dirac equations in terms of a mass parameter and two complex valued functions called ψ_1 and ψ_2. In Euclidean space in two dimensions, the Dirac equation can be written in terms of the set of Pauli matrices $\{\sigma_\mu\}$ as follows:

$$\Psi = i\left(\sigma_1 \partial_x + \sigma_2 \partial_y\right)\Psi + m\Psi = 0. \tag{1}$$

In (1), the mass term m has been generalized to be a real function of x and y, which are the Cartesian coordinates of the space. Let us introduce two complex operators defined to be

$$\partial = \frac{1}{2}\left(\partial_x - i\partial_y\right), \qquad \bar{\partial} = \frac{1}{2}\left(\partial_x + i\partial_y\right). \tag{2}$$

In terms of a complex variable $z = x + iy$, we also define $\partial = \partial/\partial z$ and $\bar{\partial} = \partial/\partial\bar{z}$. A, and spinor wavefunction Ψ is specified in terms of two components ψ_1 and ψ_2 as

$$\Psi = \begin{pmatrix} \psi_1 \\ \psi_2 \end{pmatrix}. \tag{3}$$

Using (2), the Dirac equation can be developed in terms of the two components of Ψ and their complex conjugates to give the following coupled first-order system of equations:

$$\begin{aligned} \bar{\partial}\psi_1 &= \frac{i}{2} m\psi_2, \qquad \partial\bar{\psi}_1 = -\frac{i}{2} m\bar{\psi}_2, \\ \partial\psi_2 &= \frac{i}{2} m\psi_1, \qquad \bar{\partial}\bar{\psi}_2 = -\frac{i}{2} m\bar{\psi}_1. \end{aligned} \tag{4}$$

The Dirac equation in the form (4) leads to a variety of differential constraints. The first of which is given by

$$\psi_1\partial\bar{\psi}_1 + \bar{\psi}_1\partial\psi_2 = \psi_1\left(-\frac{i}{2}m\bar{\psi}_2\right) + \bar{\psi}_2\left(\frac{i}{2}m\psi_1\right) = 0,$$

as well as its complex conjugate equation. There is also the expression for a new real variable p

$$\bar{\psi}_1\partial\psi_2 - \psi_2\partial\bar{\psi}_1 = \frac{i}{2}m\left(|\psi_1|^2 + |\psi_2|^2\right) = \frac{i}{2}mp,$$

and its complex conjugate. This also serves to define the real function P

$$p = |\psi_1|^2 + |\psi_2|^2. \tag{5}$$

A system of conservation laws can also be formulated

$$\bar{\psi}_1\partial\bar{\psi}_1 - \bar{\psi}_2\bar{\partial}\,\bar{\psi}_2 = 0, \qquad \bar{\psi}_1\partial\psi_2 + \psi_1\bar{\partial}\,\bar{\psi}_2 = 0, \tag{6}$$

as well as their complex conjugate equations. The complex quantity S is defined as follows:

$$\bar{\psi}_2\partial\psi_1 - \psi_1\partial\bar{\psi}_2 = \frac{i}{2}pS, \quad \bar{\psi}_1\bar{\partial}\psi_2 - \psi_2\bar{\partial}\psi_1 = \frac{i}{2}p\bar{S}. \tag{7}$$

Let Φ be the two-by-two matrix spinor given by

$$\Phi = \begin{pmatrix} \psi_1 & -\bar{\psi}_2 \\ \psi_2 & \bar{\psi}_1 \end{pmatrix}, \tag{8}$$

defining the real variable u so that p in (5) is given by $p = e^u$, and it follows that

$$p = e^u = \det \Phi = |\psi_1|^2 + |\psi_2|^2. \tag{9}$$

Clearly, we have $\Phi\, \Phi^\dagger = p\mathbf{I}$ and there follows another differential constraint:

$$p \partial u = \overline{\psi}_1 \partial \psi_1 + \psi_2 \partial \overline{\psi}_2. \tag{10}$$

Differentiating (8) exteriorly, we obtain that

$$d\Phi = \begin{pmatrix} \partial \psi_1 & -\partial \overline{\psi}_2 \\ \partial \psi_2 & \partial \psi_1 \end{pmatrix} dz + \begin{pmatrix} \bar{\partial} \psi_1 & -\bar{\partial}\, \overline{\psi}_2 \\ \bar{\partial} \psi_2 & \bar{\partial}\, \overline{\psi}_1 \end{pmatrix} d\bar{z}. \tag{11}$$

Consequently, we find that

$$\begin{aligned} d\Phi \cdot \Phi^{-1} &= \frac{1}{p} \begin{pmatrix} \partial \psi_1 & -\partial \overline{\psi}_2 \\ \partial \psi_2 & \partial \psi_1 \end{pmatrix} \begin{pmatrix} \overline{\psi}_1 & \overline{\psi}_2 \\ -\psi_2 & \psi_1 \end{pmatrix} dz + \frac{1}{p} \begin{pmatrix} \bar{\partial} \psi_1 & -\bar{\partial}\, \overline{\psi}_2 \\ \bar{\partial} \psi_2 & \bar{\partial}\, \overline{\psi}_1 \end{pmatrix} \begin{pmatrix} \overline{\psi}_1 & \overline{\psi}_2 \\ -\psi_2 & \psi_1 \end{pmatrix} d\bar{z} \\ &= \frac{1}{2} \left[\begin{pmatrix} 2\partial u & iS \\ im & 0 \end{pmatrix} dz + \begin{pmatrix} 0 & im \\ i\overline{S} & 2\bar{\partial} u \end{pmatrix} d\bar{z} \right]. \end{aligned} \tag{12}$$

Taking the derivative $\bar{\partial}$ of pS in (7) and substituting system (4), we obtain that

$$\begin{aligned} \bar{\partial}(pS) &= -m \overline{\psi}_1 \partial \psi_1 + \overline{\psi}_2 \partial (m \psi_2) - m \psi_2 \partial \overline{\psi}_2 + \psi_1 \partial (m \overline{\psi}_1) \\ &= -m \overline{\psi}_1 \partial \psi_1 - m \psi_2 \partial \overline{\psi}_2 + p \partial m. \end{aligned}$$

It follows that

$$p^{-1} \partial (pS) = -m p^{-1} \partial p + \partial m = p \partial (p^{-1} m) \;.$$

Let us summarize this as

$$p^{-1} \bar{\partial} (pS) = p \partial (p^{-1} m). \tag{13}$$

Proceeding in a similar fashion, we calculate the following two derivatives:

$$\begin{aligned} \partial (p^{-1} \psi_1) &= \frac{1}{p^2} \left(-|\psi_1|^2 \partial \psi_1 - \psi_1 \psi_2 \partial \overline{\psi}_2 + \left(|\psi_1|^2 + |\psi_2|^2 \right) \partial \psi_1 \right) \\ &= \frac{\psi_2}{p^2} \left(-\psi_1 \partial \overline{\psi}_2 + \overline{\psi}_2 \partial \psi_1 \right) = \frac{i}{2} S (p^{-1} \psi_2), \end{aligned} \tag{14}$$

and as well, we have

$$\begin{aligned}\bar{\partial}(p^{-1}\psi_2) &= \frac{1}{p^2}\left(-\psi_1\psi_2\overline{\partial\psi}_1 - |\psi_2|^2\bar{\partial}\psi_2 + |\psi_1|^2\bar{\partial}\psi_2 + |\psi_2|^2\bar{\partial}\psi_2\right);\\ &= \frac{\psi_1}{p}\left(-\psi_2\bar{\partial}\bar{\psi}_1 + \bar{\psi}_1\bar{\partial}\psi_2\right) = \frac{i}{2}\bar{S}(p^{-1}\psi_1).\end{aligned} \tag{15}$$

It should be pointed out that the systems (14) and (15) are summarized here

$$\partial(p^{-1}\psi_1) = \frac{i}{2}S(p^{-1}\psi_2), \qquad \bar{\partial}(p^{-1}\psi_2) = \frac{i}{2}\bar{S}(p^{-1}\psi_1). \tag{16}$$

By comparing with (4), it look very much like a Dirac system in their own right if S is thought of as a mass variable. Another quantity, a current, was found in [20] and has the form

$$J = pS.$$

It is possible to construct a vector representation of Φ as well. A matrix such as Φ represents a rotation matrix multiplied by a scaling in $\mathbb{R}^3$ as follows $V = v^i\sigma_i \to V' = \Phi V\Phi^\dagger$. So the matrix Φ can be represented by means of a multiple of an orthogona1 3×3 real matrix. The matrix elements can be found by using the inner product in V, namely $\langle V_1, V_2\rangle = (1/2)$ Tr $[V_1V_2]$, then

$$\varsigma_i^j = \frac{1}{2}Tr\left[\sigma_i\Phi\sigma_j\Phi^\dagger\right]. \tag{17}$$

ς_i^j defines a 3×3 matrix which can be written down by using the usual representation of the Pauli matrices. In particular, the matrix formed out of the following combinations will be very useful:

$$\begin{aligned}\varsigma_+ &= \frac{1}{\sqrt{2}}(\varsigma_1 - i\varsigma_2) = \frac{1}{\sqrt{2}}\left(\psi_1^2 - \bar{\psi}_2^2, \quad -i\left(\psi_1^2 + \bar{\psi}_2^2\right), 2\psi_1\bar{\psi}_2\right),\\ \varsigma_- &= \frac{1}{\sqrt{2}}(\varsigma_1 + i\varsigma_2) = \frac{1}{\sqrt{2}}\left(\bar{\psi}_1^2 - \psi_2^2, \ i\left(\bar{\psi}_1^2 + \psi_2^2\right), 2\bar{\psi}_1\psi_2\right),\\ \varsigma_3 &= \left(-\psi_1\psi_2 - \bar{\psi}_1\bar{\psi}_2, \ i(\psi_1\psi_2 - \bar{\psi}_1\bar{\psi}_2), \ |\psi_1|^2 - |\psi_2|^2\right).\end{aligned} \tag{18}$$

In terms of matrices, ς and $\varsigma^\dagger$ are represented as:

$$\varsigma = \begin{pmatrix} \frac{1}{\sqrt{2}}\left(\psi_1^2 - \bar{\psi}_2^2\right) & -\frac{i}{\sqrt{2}}\left(\psi_1^2 + \bar{\psi}_2^2\right) & 2\psi_1\bar{\psi}_2 \\ \frac{1}{\sqrt{2}}\left(\bar{\psi}_1^2 - \psi_2^2\right) & \frac{i}{\sqrt{2}}\left(\bar{\psi}_1^2 + \psi_2^2\right) & 2\bar{\psi}_1\psi_2 \\ -\psi_1\psi_2 - \bar{\psi}_1\psi_2 & i(\psi_1\psi_2 - \bar{\psi}_1\bar{\psi}_2) & |\psi_1|^2 - |\psi_2|^2 \end{pmatrix}, \tag{19}$$

$$
\varsigma\dagger = \begin{pmatrix} \frac{1}{\sqrt{2}}\left(\overline{\psi}_1^2 - \overline{\psi}_2^2\right) & \frac{1}{2}\left(\psi_1^2 + \overline{\psi}_2^2\right) & -(\overline{\psi}_1\overline{\psi}_2 + \psi_1\psi_2) \\ \frac{i}{\sqrt{2}}\left(\overline{\psi}_1^2 - \psi_2^2\right) & -\frac{i}{\sqrt{2}}\left(\psi_1^2 + \overline{\psi}_2^2\right) & i(\psi_1\psi_2 - \overline{\psi}_1\overline{\psi}_2) \\ \sqrt{2}\overline{\psi}_1\psi_2 & \sqrt{2}\psi_1\overline{\psi}_2 & |\psi_1|^2 - |\psi_2|^2 \end{pmatrix}
$$

Given this explicit representation, it is now possible to evaluate

$$
p^{-2}d\varsigma \cdot \varsigma^{\dagger} = p^{-2}\left(\partial\varsigma \cdot \varsigma^{\dagger}dz + \overline{\partial}\varsigma \cdot \varsigma^{\dagger}d\overline{z}\right). \tag{20}
$$

To obtain an expression for (20), both matrices (19) can be expressed in Maple. Apply the operator $map(\partial)$ to ς, right multiply by $\varsigma^{\dagger}$ then substitute system (4) of known derivatives to obtain the matrix

$$
\begin{pmatrix} 2(\psi_2\partial\overline{\psi}_2 + \overline{\psi}_1\partial\psi_1)p & 0 & \sqrt{2}(\psi_1\partial\overline{\psi}_2 - \overline{\psi}_2\partial\psi_1)p \\ 0 & 0 & -\frac{i}{\sqrt{2}}mp^2 \\ -\frac{i}{\sqrt{2}}mp^2 & \sqrt{2}(\overline{\psi}_2\partial\overline{\psi}_1 - \psi_1\partial\overline{\psi}_2)p & (\psi_2\partial\overline{\psi}_2 + \overline{\psi}_1\partial\psi_1)p \end{pmatrix} \tag{21}
$$

Similarly, applying $map(\overline{\partial})$ to ς then right multiplying by $\varsigma^{\dagger}$ yields

$$
\begin{pmatrix} 0 & 0 & -\frac{i}{\sqrt{2}}mp^2 \\ 0 & 2(\psi_1\overline{\partial}\psi_1 + \overline{\psi}_2\overline{\partial}\psi_2)p & -\sqrt{2}(\psi_2\overline{\partial}\,\overline{\psi}_1 - \overline{\psi}_1\overline{\partial}\psi_2)p \\ \frac{i}{\sqrt{2}}p^2\overline{S} & \frac{i}{\sqrt{2}}mp^2 & (\psi_1\overline{\partial}\,\overline{\psi}_1 + \psi_2\overline{\partial}\psi_2)p \end{pmatrix} \tag{22}
$$

By (20) and the differential constraints, the vector representation of the Maurer-Cartan form can be expressed as:

$$
p^{-2}d\varsigma\varsigma^{\dagger} = \begin{pmatrix} 2\partial u & 0 & -\frac{i}{\sqrt{2}}S \\ 0 & 0 & \frac{i}{2}m \\ -\frac{i}{\sqrt{2}}m & \frac{i}{\sqrt{2}}S & \partial u \end{pmatrix} dz + \begin{pmatrix} 0 & 0 & -\frac{i}{\sqrt{2}}m \\ 0 & 2\overline{\partial}u & \frac{i}{\sqrt{2}}\overline{S} \\ -\frac{i}{\sqrt{2}}\overline{S} & \frac{i}{\sqrt{2}}m & \overline{\partial}u \end{pmatrix} dz. \tag{23}
$$

According to the properties of the inner product, we can write $E_i = \varsigma_i^j \sigma_j = \Phi^\dagger \sigma_i \Phi$ and calculate that

$$d\varsigma_i \varsigma_j = \langle dE_i, E_j \rangle = \frac{1}{2}\,\mathrm{Tr}\left[d\left(\Phi^\dagger \sigma_i \Phi\right)\Phi^\dagger \sigma_j \Phi\right] = \frac{1}{2}\,\mathrm{Tr}\left[\left(d\Phi^\dagger \sigma_i \sigma_j \Phi\right) + \left(d\Phi^\dagger \sigma_i \sigma_j \Phi\right)^\dagger\right]. \tag{24}$$

If a conserved current can be constructed whose components are divergence free, then a differential one-form exists with values in $\mathbb{R}^3$ that will induce a surface upon quadrature. Such a current will be given from the global symmetries of the Lagrangian by means of Noether's theorem. Making the transformations $\psi_1 \to -\overline{\psi}_2$ and $\psi_2 \to \overline{\psi}_1$ in system (4), it is seen to remain invariant. This can be thought of as a charge conjugation. The same solutions are obtained if we put Φ instead of Ψ in the Dirac equation (2). So Φ multiplied on the right by any constant nonsingular matrix is a solution of the equation if Φ is. This implies the full symmetry group is $GL(2\mathbb{C})$. The transformation above is a member of this group, so can be thought of as a continuous transformation. In terms of matrix Φ, the Lagrangian of the Dirac equation can be written as

$$\mathcal{L} = \frac{1}{2}\mathrm{Tr}\left[\Phi^\dagger \Phi\right] = i\overline{\psi}_1 \partial \psi_2 - i\psi_2 \partial \overline{\psi}_1 + i\overline{\psi}_2 \overline{\partial} \psi_1 - i\psi_1 \overline{\partial}\,\overline{\psi}_2 + m\left(|\psi_1|^2 + |\psi_2|^2\right). \tag{25}$$

The currents that correspond to the generators of $SU(2)$ are found to be proportional to the components of ς_+ and ς_-; hence, the required conservation law is

$$\partial \varsigma_- + \overline{\partial} \varsigma_+ = 0. \tag{26}$$

Alternatively, the Dirac equation and its Hermitian conjugate which are given by

$$\Phi^\dagger\left(i\sigma_1 \partial_x + i\sigma_2 \partial_y + m\right)\Phi = 0 \quad \Phi^\dagger\left(i\overleftarrow{\partial}_x \sigma_1 + i\overleftarrow{\partial}_y \sigma_2 - m\right)\Phi = 0, \tag{27}$$

may be added to obtain

$$\partial_x\left(\Phi^\dagger \sigma_1 \Phi\right) + \partial_y\left(\Phi^\dagger \sigma_2 \Phi\right) = 0. \tag{28}$$

Now to describe the surface, define the $\mathbb{R}^3$-valued differential form

$$d\mathbf{r} = \frac{i}{\sqrt{2}}\varsigma_+\, dz + \frac{i}{\sqrt{2}}\varsigma_- d\overline{z} \tag{29}$$

which is real since $\varsigma_- = \overline{\varsigma}_+$. The differential form (29) is closed under substitution of conservation law (26) since

$$d^2\mathbf{r} = \frac{i}{\sqrt{2}}\overline{\partial}\varsigma_+\, d\overline{z} \wedge dz - \frac{i}{\sqrt{2}}\partial \varsigma_- dz \wedge d\overline{z} = \frac{i}{\sqrt{2}}\left(\overline{\partial}\varsigma_+ + \partial\varsigma_-\right) dz \wedge d\overline{z} = 0. \tag{30}$$

By Poincare's lemma, the form is exact since every loop in $\mathbb{C}$ can be collapsed to a point. Therefore, the desired expression for a surface will result when the form is integrated along a path Γ^1 in the $(z, \overline{z})$ plane from a fixed point z_0. The components are

$$\begin{aligned} dx_1 &= \frac{i}{2}\left(\psi_1^2 - \overline{\psi}_2^2\right)dz - \frac{i}{2}\left(\overline{\psi}_1^2 - \psi_2^2\right)d\overline{z}, \\ dx_2 &= \frac{1}{2}\left(\psi_1^2 + \overline{\psi}_2^2\right)dz + \frac{1}{2}\left(\overline{\psi}_1^2 + \psi_2^2\right)d\overline{z}, \\ dx_3 &= i(\psi_1\overline{\psi}_2 dz - \overline{\psi}_1\psi_2 d\overline{z}) \ . \end{aligned} \tag{31}$$

Combining the first two equations in (31) and integrating from z_0, the coordinates of a surface in $\mathbb{R}^3$ are obtained by integrating over any path Γ^1 in the $(z, \overline{z})$ plane

$$\begin{aligned} x_1 + ix_2 &= i\int_\Gamma (\psi_1^2 dz' + \psi_2^2 d\overline{z}') \\ x_1 - ix_2 &= -i\int_\Gamma \left(\overline{\psi}_2^2 dz' + \overline{\psi}_1^2 d\overline{z}'\right), \\ x_3 &= i\int_\Gamma (\psi_1\overline{\psi}_2 dz' - \overline{\psi}_1\psi_2 d\overline{z}') \ . \end{aligned} \tag{32}$$

In the end, we have set z_0 to be zero, and it may be repeated; the integrals are independent of Γ due to the conservation laws. In (31) and (32), $\mathbf{r}$ is the point of the surface with coordinates $(x_1, x_2, x_3) \in \mathbb{R}^3$ and ς_3 is normal to the surface.

3. Fundamental forms and Cartan moving frame

The necessary information to write down the traditional data for a surface has been obtained. Since $\varsigma_\pm^2 = 0$ and $\varsigma_+ \cdot \varsigma_- = 0$, the first fundamental form is given by

$$I = d\mathbf{r} \cdot d\mathbf{r} = p^2 dz \otimes d\overline{z} \tag{33}$$

or in a matrix representation,

$$I = \frac{p^2}{2}\begin{pmatrix} 0 & 1 \\ 1 & 0 \end{pmatrix}.$$

The inverse of (33) is given by

$$I^{-1} = \frac{2}{p^2}\begin{pmatrix} 0 & 1 \\ 1 & 0 \end{pmatrix} \tag{34}$$

It is therefore a conformal immersions with isothermal coordinates ζ_1, ζ_2. The second fundamental form of the surface can also be calculated and using $\varsigma_3 \cdot d\mathbf{r} = 0$,

$$II = -d(p^{-1}\varsigma_3) \cdot \ d\mathbf{r} = -p^{-1} d\varsigma_3 \cdot d\mathbf{r} = -\frac{p}{2}\left(S dz \otimes dz + 2m dz \otimes d\overline{z} + \overline{S} d\overline{z} \otimes d\overline{z}\right) \tag{35}$$

and in matrix form,

$$II = -\frac{p}{2}\begin{pmatrix} S & m \\ m & \overline{S} \end{pmatrix}.$$

Collecting (34) and (35), we have

$$II \cdot I^{-1} = -\frac{p}{2}\begin{pmatrix} S & m \\ m & \overline{S} \end{pmatrix}\begin{pmatrix} 0 & \frac{2}{p^2} \\ \frac{2}{p^2} & 0 \end{pmatrix} = -\frac{1}{p}\begin{pmatrix} S & m \\ m & \overline{S} \end{pmatrix}\begin{pmatrix} 0 & 1 \\ 1 & 0 \end{pmatrix} = -\frac{1}{p}\begin{pmatrix} m & S \\ \overline{S} & m \end{pmatrix}.$$

The usual definitions give the mean curvature H and the Gaussian curvature as well

$$H = \frac{1}{2}\mathrm{Tr}\left(II \cdot I^{-1}\right) = -\frac{m}{p}, \tag{36}$$

$$K = \det\left(II \cdot I^{-1}\right) = \frac{1}{p^2}\left(m^2 - |S|^2\right) = H^2 - \frac{|S|^2}{p^2}. \tag{37}$$

Equation (36) relates the mean curvature H to the mass parameter in the Dirac equation. Konopelchenko obtains the expression

$$K = -4p^{-2}\partial\overline{\partial}u, \tag{38}$$

which is known as the Gauss-Riemann curvature. It has been shown however that it is equivalent to (37) in accord with Gauss' Theorem Egregium.

It is interesting to note that since the difference between the principal curvatures is given as

$$(\Delta\kappa)^2 = 4\left(H^2 - K\right), \tag{39}$$

it also holds that since $H^2 - K = p^{-2}|S|^2$,

$$|\Delta\kappa| = \frac{2}{p}|S|.$$

Thus, the modulus of S is a measure of the local deformation from a spherical surface as m is a measure of the local deformation from the case of a minimal surface, so $\kappa = -p^{-1}(m \pm |S|)$.

A fixed referential frame in $\mathbb{R}^3$ has been implicitly used up to now. By varying the frame with some solution of Dirac system (4), a whole set of surfaces is obtained that may be deduced from each other by means of a rigid motion. Cartan developed a powerful method referred to as the moving frame method to avoid this awkward process.

By introducing differential 1-forms also called Pfaffian forms, we define the system

$$d\mathbf{r} = \omega^j \mathbf{e}_j, \qquad d\mathbf{e}_j = \omega_i^j \mathbf{e}_j, \qquad i, j = +, -, 3. \tag{40}$$

This is the first system of structure equations introduced by Cartan. The vectors $\mathbf{e}_i$ satisfy orthonormality conditions

$$\mathbf{e}_+^2 = 0, \qquad \mathbf{e}_+ \cdot \mathbf{e}_- = p^2, \qquad \mathbf{e}_\pm \cdot \mathbf{e}_3 = 0, \qquad \mathbf{e}_3^2 = p^2. \tag{41}$$

Differentiating relations (41) and using structure equations (40), the following relations among the differential forms are obtained

$$d\mathbf{e}_+^2 = 0, \qquad 2\mathbf{e}_+ \cdot d\mathbf{e}_+ = 0, \qquad 2\mathbf{e}_+ \left(\omega_+^j \mathbf{e}_j\right) = 0,$$

$$\begin{gathered}
2\mathbf{e}_+ \cdot \omega_+^- \mathbf{e}_- = 0 \quad \omega_+^- = 0. \qquad \square \\
d\mathbf{e}_-^2 = 0, \qquad 2\mathbf{e}_- \cdot d\mathbf{e}_- = 0, \qquad 2\mathbf{e}_- \cdot (\omega_-^j \mathbf{e}_j) = 0, \\
2\mathbf{e}_- \ (\omega_-^+ \mathbf{e}_+) = 0, \quad \omega_-^+ = 0. \quad \square \\
d\mathbf{e}_+ \cdot \mathbf{e}_- + \mathbf{e}_+ \cdot d\mathbf{e}_- = 2pdp, \\
\omega_+^+ \mathbf{e}_+ \cdot \mathbf{e}_- + \omega_-^- \mathbf{e}_+ \cdot \mathbf{e}_- = 2pdp, \\
\omega_-^- + \omega_+^+ = 2\partial u, \qquad \square \\
d\mathbf{e}_+ \cdot \mathbf{e}_3 + \mathbf{e}_+ \cdot d\mathbf{e}_3 = 0, \\
\omega_j^3 + \omega_3^- = 0. \quad \square \\
d\mathbf{e}_- \cdot \mathbf{e}_3 + \mathbf{e}_- \omega_3^j \mathbf{e}_j = 0, \\
\omega_-^3 + \omega_3^+ = 0. \quad \square \\
3\mathbf{e}_3 \cdot d\mathbf{e}_3 = p^2 du \qquad \omega_3^3 = du. \qquad \square
\end{gathered}$$

This collection of results is summarized all together below

$$\begin{gathered}
\omega_-^+ = \omega_+^- = 0, \\
\omega_3^- + \omega_+^3 = \omega_3^+ + \omega_-^3 = 0, \\
\omega_-^- + \omega_+^+ = 2du, \qquad \omega_3^3 = du.
\end{gathered} \tag{42}$$

As $\overline{\mathbf{e}}_+ = \mathbf{e}_-$ and $\overline{\mathbf{e}}_3 = \mathbf{e}_3$, it is found that

$$\overline{\omega}_+^3 = \omega_-^3, \qquad \overline{\omega}_+^+ = \omega_-^-. \tag{43}$$

Assuming structure equations (40) are integrable, differentiating and substituting $d\mathbf{e}_i$ where ever possible, compatibility equations are obtained which are referred to as the second system of structure equations, that is first we have

$$d\omega^j \mathbf{e}_j - \omega^j \wedge d\mathbf{e}_j = 0,$$

hence

$$d\omega^s = \omega^j \wedge \omega_j^s,$$

and next

$$d\omega_i^j \mathbf{e}_j - \omega_i^j \wedge d\mathbf{e}_j = 0,$$

hence

$$d\omega_i^j = \omega_i^j \wedge \omega_j^s.$$

Let us summarize these as the pair

$$d\omega^s = \omega^j \wedge \omega_j^s, \qquad d\omega_i^s = \omega_i^j \wedge \omega_j^s. \tag{44}$$

The second equality is always true as long as the frames are given, and the first is the equivalent, expressed in the formalism of a moving frame, of the requirement that the form $d\mathbf{r}$ be exact. Writing $d\mathbf{r}$ as

$$d\mathbf{r} = \frac{i}{\sqrt{2}}\mathbf{e}_+ dz - \frac{i}{\sqrt{2}}\mathbf{e}_- d\bar{z} = \omega^j \mathbf{e}_j. \tag{45}$$

Let us identify the forms

$$\omega^+ = \frac{i}{\sqrt{2}} dz, \qquad \omega^- = -\frac{i}{\sqrt{2}} d\bar{z}, \qquad \omega^3 = 0, \ \overline{\omega}^+ = \omega^-. \tag{46}$$

The equations for the remaining one-forms can be represented by writing the structure equation in the form

$$d\mathbf{e} = \Omega \mathbf{e}. \tag{47}$$

In (47), Ω is represented by the 3×3 matrix of forms

$$\Omega = \begin{pmatrix} \omega_+^+ & \omega_+^- & \omega_+^3 \\ \omega_-^+ & \omega_-^- & \omega_-^3 \\ \omega_3^+ & \omega_3^- & \omega_3^3 \end{pmatrix} \tag{48}$$

Since $\mathbf{e} \cdot \mathbf{e}^\dagger = p^2 \mathbf{I}$, (47) can be right multiplied by $\mathbf{e}^\dagger$ to obtain

$$\Omega = p^{-2} \ d\mathbf{e} \cdot \mathbf{e}^\dagger. \tag{49}$$

This implies that Ω can be identified with the Maurer-Cartan form given in (22). Introduce the vector of differential forms Θ as

$$\Theta = (\omega^+, \ \omega^-, \ 0), \quad d\Theta = 0. \tag{50}$$

In terms of Θ the compatibility equations take the form

$$d\Theta = \Theta \wedge \Omega, \qquad d\Omega = \Omega \wedge \Omega. \tag{51}$$

It is clear from the Maurer-Cartan form that it can be decomposed in the following manner

$$\Omega = M_1 dz + M_2 d\bar{z}, \tag{52}$$

where M_1 and M_2 are defined to be the matrices

$$M_1 \begin{pmatrix} 2\,\partial u & 0 & -\frac{i}{\sqrt{2}}S \\ 0 & 0 & \frac{i}{\sqrt{2}}m \\ -\frac{i}{\sqrt{2}}m & \frac{i}{\sqrt{2}}S & \partial u \end{pmatrix}, \qquad M_2 = \begin{pmatrix} 0 & 0 & -\frac{i}{\sqrt{2}}m \\ 0 & 2\bar{\partial}u & \frac{i}{\sqrt{2}}\bar{S} \\ -\frac{\bar{S}}{\sqrt{2}} & i\frac{m}{\sqrt{2}} & \bar{\partial}u \end{pmatrix}. \tag{53}$$

The first structure equation in (33) is then

$$\partial e = M_1 e, \qquad \bar{\partial} e = M_2 e. \tag{54}$$

This corresponds to the Gauss-Weingarten equation and the second compatibility equation

$$\bar{\partial} M_1 - \partial M_2 + [M_1, M_2] = 0, \tag{55}$$

is also known as the Gauss-Codazzi-Mainardi equations. All of these have been seen here before in (37) and (45). It has been shown that many nonlinear partial differential equations can be expressed within this formalism. In a spinor representation, the corresponding representation in the form of matrices can be obtained out of the Maurer-Cartan form

$$Z_1 = \frac{1}{2}\begin{pmatrix} 2\partial u & iS \\ im & 0 \end{pmatrix}, \qquad Z_2 = \frac{1}{2}\begin{pmatrix} 0 & im \\ i\bar{S} & 2\bar{\partial}u \end{pmatrix}. \tag{56}$$

In terms of these matrices, the linear system is

$$\partial \Phi = Z_1 \Phi, \qquad \bar{\partial}\Phi = Z_2 \Phi,$$
$$\bar{\partial} Z_1 - \partial Z_2 + [Z_1,\ Z_2] = 0. \tag{57}$$

The differential form Ω is a connection and actually an affine connection on $\mathbb{R}^3$. It is flat on the surface. This is the meaning of the second system of structure equations. This means that the two-dimensional Dirac equation can be regarded as a way of expressing an affine connection.

Two make further progress, Ω can be used in the following way. As $\omega^3 = 0$, from the compatibility equation for $d\omega^3$, we have

$$\omega^+ \wedge \omega^3_+ + \omega^- \wedge \omega^3_- = 0.$$

On account of Cartan's lemma, both ω^3_+ and ω^3_- are equal to a linear combination of ω^+ and ω^-

$$\omega^3_+ = h_{++}\omega^+ + h_{+-}\omega^-, \qquad \omega^3_- = h_{-+}\omega^+ + h_{--}\omega^-, \tag{58}$$

where $h_{-+} = h_{+-}$, as can be seen by substituting ω_+^3 and ω_-^3 into the constraint above. Since

$$\Omega = \begin{pmatrix} 2\partial u & 0 & -\frac{i}{\sqrt{2}}S \\ 0 & 0 & \frac{i}{\sqrt{2}}m \\ -\frac{i}{\sqrt{2}}m & \frac{i}{\sqrt{2}}S & \partial u \end{pmatrix} dz + \begin{pmatrix} 0 & 0 & -\frac{i}{\sqrt{2}}m \\ 0 & \bar{\partial} u & \frac{i}{\sqrt{2}}\overline{S} \\ -\frac{i}{\sqrt{2}}\overline{S} & \frac{i}{\sqrt{2}}m & \bar{\partial} u \end{pmatrix} d\bar{z}. \tag{59}$$

Using $\omega^+, \ \omega^-$ and ω^3 from (46), we have

$$\omega_+^3 = h_{++}\left(\frac{i}{\sqrt{2}}\right)dz + h_{+-}\left(-\frac{i}{\sqrt{2}}\right)d\bar{z} = -\frac{i}{\sqrt{2}}Sdz - \frac{i}{\sqrt{2}}md\bar{z}. \tag{60}$$

This relation implies that

$$h_{++} = -S, \qquad h_{+-} = -m, \tag{61}$$

and moreover, it follows that

$$\omega_-^3 = h_{-+}\omega^+ + h_{--}\omega^- = h_{-+}\left(\frac{i}{\sqrt{2}}\right)dz + h_{--}\left(-\frac{i}{\sqrt{2}}\right)d\bar{z} = \frac{i}{\sqrt{2}}mdz + \frac{i}{\sqrt{2}}\overline{S}d\bar{z}.$$

This implies that

$$h_{-+} = m, \qquad h_{--} = -\overline{S}. \tag{62}$$

It is important to note that these coefficients can be used together with the structure equations to express the fundamental forms of the surface in terms of Pfaffian forms. The first fundamental form is given as

$$I = 2p^2\omega^+ \otimes \omega^- = 2p^2\left(\frac{1}{2}\right)dz \otimes d\bar{z}, \tag{63}$$

and the second fundamental form can be written as

$$\begin{aligned} II &= -p\left(\omega^+ \otimes \omega_+^3 + \omega^- \otimes \omega_-^3\right) \\ &= -p(h_{++}\omega^+ \otimes \omega^+ + (h_{+-} + h_{-+})\omega^+ \otimes \omega^- + h_{--}\omega^- \otimes \omega^-). \end{aligned} \tag{64}$$

The element of surface is given by

$$dS = ip^2\omega^+ \wedge \omega^-, \tag{65}$$

and the corresponding surface element on the Gauss map is

$$d\sigma = i\omega_-^3 \wedge \omega_+^3 = i(h_{+-}h_{-+} - h_{++}h_{--})\omega^+ \wedge \omega^-. \tag{66}$$

The total curvature would be the ratio of the former to the latter,

$$K = p^{-2}(h_{+-}h_{-+} - h_{++}h_{--}). \tag{67}$$

Finally, the mean curvature is given as

$$H = -\frac{1}{2p}(h_{+-} + h_{-+}). \tag{68}$$

4. The Gauss map and nonlinear Sigma model

Under the condition that a given moving frame is integrable, the surface is defined up to a translation. Conversely, given the three vectors which constitute the frame, only one is determined uniquely by the surface, and that is the normal vector. For this reason, it is often referred to as the Gauss or spherical map, as it maps the parameter plane to the sphere of radius one in two dimensions. The map in this instance is given as

$$\phi = \frac{\mathbf{e}_3}{p}, \tag{69}$$

so the north pole corresponds to $\psi_2 = 0$, while the south pole to $\psi_1 = 0$. If the first column of $\Phi^\dagger$ is considered as well as the associated fundamental field

$$\rho = -\frac{\psi_2}{\psi_1}, \tag{70}$$

then dividing the numerator and denominator by $|\psi_1|^2$ in (69), we obtain

$$\phi = \frac{1}{1+|\rho|^2}\left(\rho + \bar{\rho},\; i(\bar{\rho} - \rho), 1 - |\rho|^2\right). \tag{71}$$

This quantity is a function of only ρ. It may be thought that ρ plays the role of stereographic projection of the Gauss map from the south pole. Moreover, for a minimal surface where $m = 0$, it is readily shown that $\bar{\rho}$ is an analytic function of z.

Using the differential constraints, the derivatives of ρ are found to be

$$\partial\rho = -ip\frac{m}{2\bar{\psi}_1^2}, \qquad \partial\bar{\rho} = ip\frac{S}{2\psi_1^2}. \tag{72}$$

By using derivatives (72), the following three relations can be worked out

$$4\frac{\partial\rho\bar{\partial}\bar{\rho}}{\left(1+|\rho|^2\right)^2} = m^2, \qquad 4\frac{\partial\rho\partial\bar{\rho}}{\left(1+|\rho|^2\right)^2} = mS, \qquad 4\frac{\bar{\partial}\rho\partial\bar{\rho}}{\left(1+|\rho|^2\right)^2} = |S|^2. \tag{73}$$

Thus, the quantities m and S can be written as a function of only ρ. It may be asked, can the component of the Maurer-Cartan form ∂u be written in a similar way? Starting with the differential constraint for ∂u,

$$\partial u = \frac{1}{p}\left(\overline{\psi}_1 \partial \psi_1 + \psi_2 \partial \overline{\psi}_2\right) = e^{-u/2} \partial\left(\psi_1, -\overline{\psi}_2\right) e^{-u/2} \begin{pmatrix} \overline{\psi}_1 \\ \psi_2 \end{pmatrix}. \tag{74}$$

Since the spinor product $\left(\psi_1, -\overline{\psi}_2\right)\left(\psi_1, -\overline{\psi}_2\right)^{\dagger} = e^u$, we have

$$\partial\left[e^{-u/2}\left(\psi_1, -\overline{\psi}_2\right)\right] e^{-u/2} \begin{pmatrix} \overline{\psi}_1 \\ -\psi_2 \end{pmatrix} = -\frac{1}{2} \partial u \left(\psi_1, -\overline{\psi}_2\right) e^{-u} \begin{pmatrix} \overline{\psi}_1 \\ -\psi_2 \end{pmatrix} + e^{-u/2} \partial\left(\psi_1, -\overline{\psi}_2\right) e^{-u/2} \begin{pmatrix} \overline{\psi}_1 \\ -\psi_2 \end{pmatrix}.$$

Combining these last two results, we obtain

$$\frac{1}{2} \partial u = \partial\left[e^{-u/2}\left(\psi_1, -\overline{\psi}_2\right)\right] e^{-u/2} \begin{pmatrix} \overline{\psi}_1 \\ -\psi_2 \end{pmatrix}. \tag{75}$$

If we define the spinor $\alpha = e^{-u/2}\left(\psi_1, -\overline{\psi}_2\right)$ which satisfies $\alpha\ \alpha^{\dagger} = \mathbf{1}$, (75) becomes

$$\frac{1}{2} \partial u = \partial \alpha \alpha^{\dagger}. \tag{76}$$

Let us show that α can be expressed as a function of ρ. Using the definition of ρ, a parameterization for α exists as

$$\alpha = \frac{1}{\sqrt{1 + |\rho|^2}} (1, \overline{\rho}) \left(\frac{\psi_1}{\overline{\psi}_2}\right)^{1/2}. \tag{77}$$

To obtain an expression for ψ_1, use differential constraint (2) its conjugate and (70) to arrive at

$$\overline{\psi}_1^2 \partial \rho = -\frac{i}{2} p m. \tag{78}$$

Dividing this by its complex conjugate gives α as a function of ρ as

$$\alpha = \frac{1}{\sqrt{1 + |\rho|^2}} (1,\ \rho) \left(\frac{\partial \rho}{\overline{\partial} \overline{\rho}}\right)^{1/4} e^{i(2n+1)\pi/4}, \qquad n \in \mathrm{Z}. \tag{79}$$

Inserting α into the expression for ∂u provides expressions for ∂u and $\overline{\partial} u$

$$\partial\left(\frac{\partial \rho}{\overline{\partial} \overline{\rho}}\right) = 2 \log\left(\frac{\partial \rho}{\overline{\partial} \overline{\rho}}\right) \cdot \frac{\partial \rho}{\overline{\partial} \overline{\rho}}. \tag{80}$$

Differentiating the components of α with respect to z, we find that

$$\begin{aligned}\partial\alpha\alpha^{\dagger} &= -\frac{(\bar{\rho}\partial\rho + \rho\partial\bar{\rho})}{2\left(1+|\rho|^2\right)^2} + \frac{1}{4\left(1+|\rho|^2\right)}\partial\log\left(\frac{\partial\rho}{\bar{\partial}\bar{\rho}}\right) \\ &+ \frac{-\rho\bar{\rho}^2\partial\rho - \rho^2\bar{\rho}\partial\bar{\rho} + 2\rho\left(1+|\rho|^2\right)\partial\bar{\rho}}{2\left(1+|\rho|^2\right)^2} + \frac{|\rho|^2}{4\left(1+|\rho|^2\right)}\partial\log\left(\frac{\partial\rho}{\bar{\partial}\bar{\rho}}\right) \\ &= \frac{-\bar{\rho}\partial\rho + \rho\partial\bar{\rho}}{2\left(1+|\rho|^2\right)} + \frac{1}{4}\partial\log\left(\frac{\partial\rho}{\bar{\partial}\bar{\rho}}\right).\end{aligned}$$

Returning to the expression for $(1/2)\partial u$, we can now write

$$\frac{1}{2}\partial u = \frac{1}{4}\partial\log\left(\frac{\partial\rho}{\bar{\partial}\bar{\rho}}\right) + \frac{1}{2}\frac{\rho\partial\bar{\rho} - \bar{\rho}\partial\rho}{1+|\rho|^2}, \tag{81}$$

$$\frac{1}{2}\bar{\partial}u = -\frac{1}{4}\bar{\partial}\log\left(\frac{\partial\rho}{\bar{\partial}\bar{\rho}}\right) + \frac{1}{2}\frac{\bar{\rho}\bar{\partial}\rho - \rho\bar{\partial}\bar{\rho}}{1+|\rho|^2} \tag{82}$$

There is no simple integral of the second term in general. It may be stated that $(1/2)\partial u$ has the form of a potential with a fixed gauge, because ρ is given as a function of z and $\bar{z}$, so the directions of the axes $\mathbf{e}_+$ and $\mathbf{e}_-$ have been fixed so that a gauge transformation is a rotation of them.

Suppose it is asked under what condition a given complex function $\rho(z,\bar{z})$ is the Gauss map of some surface. A necessary condition can be obtained by working out the compatibility condition for the linear system (81) and (82), that is, first

$$\begin{aligned}\partial\bar{\partial}u &= -\frac{1}{2}\partial\bar{\partial}\log\left(\frac{\partial\rho}{\bar{\partial}\bar{\rho}}\right) + \frac{\partial\bar{\rho}\,\bar{\partial}\rho + \bar{\rho}\partial\bar{\partial}\rho - \partial\rho\,\bar{\partial}\bar{\rho} - \rho\partial\,\bar{\partial}\bar{\rho}}{1+|\rho|^2} - \frac{\bar{\rho}\bar{\partial}\rho - \rho\bar{\partial}\bar{\rho}}{\left(1+|\rho|^2\right)^2}(\rho\partial\bar{\rho} + \bar{\rho}\partial\rho) \\ &= -\frac{1}{2}\partial\bar{\partial}\log\left(\frac{\partial\rho}{\bar{\partial}\bar{\rho}}\right) + \frac{\bar{\partial}\rho\partial\bar{\rho} - \bar{\partial}\bar{\rho}\partial\rho - \bar{\partial}\rho\partial\rho\bar{\rho}^2 + \partial\bar{\rho}\bar{\partial}\bar{\rho}\rho^2 - \left(1+|\rho|^2\right)(\rho\partial\bar{\partial}\bar{\rho} - \bar{\rho}\partial\bar{\partial}\rho)}{\left(1+|\rho|^2\right)^2},\end{aligned}$$

and the result for the other mixed derivative is

$$\bar{\partial}\partial u = \frac{1}{2}\bar{\partial}\partial\log\left(\frac{\partial\rho}{\bar{\partial}\bar{\rho}}\right) + \frac{\partial\bar{\rho}\bar{\partial}\rho - \bar{\partial}\bar{\rho}\partial\rho + \left(1+|\rho|^2\right)(\rho\partial\bar{\partial}\bar{\rho} - \bar{\rho}\partial\bar{\partial}\rho) + \bar{\rho}^2\partial\rho\bar{\partial}\rho - \rho^2\partial\bar{\rho}\bar{\partial}\bar{\rho}}{\left(1+|\rho|^2\right)^2}.$$

Equating these mixed partial derivatives, the necessary condition takes the form

$$\partial\bar{\partial}\log\left(\frac{\partial\rho}{\bar{\partial}\bar{\rho}}\right) + 2\frac{\left(1+|\rho|^2\right)(\rho\partial\bar{\partial}\bar{\rho} - \bar{\rho}\partial\bar{\partial}\rho) + \bar{\rho}^2\partial\rho\bar{\partial}\rho - \rho^2\partial\bar{\rho}\bar{\partial}\bar{\rho}}{\left(1+|\rho|^2\right)^2}. \tag{83}$$

If it is satisfied, it has the implication that

$$\partial\bar{\partial}u = \frac{\bar{\partial}\rho\partial\bar{\rho} - \partial\rho\bar{\partial}\bar{\rho}}{\left(1+|\rho|^2\right)^2}. \tag{84}$$

Using the previous expressions (73) for the derivatives of ρ, this can be put into the form of the Gauss equation. Consequently, one of the integrability conditions is fulfilled. Since

$$\frac{\bar{\partial}\rho\partial\bar{\rho} - \partial\rho\bar{\partial}\bar{\rho}}{\left(1+|\rho|^2\right)^2} = \partial\left(\frac{1}{2}\frac{\bar{\rho}\bar{\partial}\rho - \rho\bar{\partial}\bar{\rho}}{1+|\rho|^2}\right) + \bar{\partial}\left(\frac{1}{2}\frac{\rho\partial\bar{\rho} - \bar{\rho}\partial\rho}{1+|\rho|^2}\right), \tag{85}$$

it follows that

$$\partial\bar{\partial}u = \partial\left(\frac{1}{2}\frac{\bar{\rho}\bar{\partial}\rho - \rho\bar{\partial}\bar{\rho}}{1+|\rho|^2}\right) + \bar{\partial}\left(\frac{1}{2}\frac{\rho\partial\bar{\rho} - \bar{\rho}\partial\rho}{1+|\rho|^2}\right) = \frac{\bar{\partial}\rho\partial\bar{\rho} - \partial\rho\bar{\partial}\bar{\rho}}{\left(1+|\rho|^2\right)^2} = \frac{1}{4}p^2K. \tag{86}$$

Due to cancelations, some shorthand expressions might be quoted

$$\partial\bar{\partial}u = \partial\left(\frac{\bar{\rho}\bar{\partial}\rho}{1+|\rho|^2}\right) + \bar{\partial}\left(-\frac{\bar{\rho}\,\partial\rho}{1+|\rho|^2}\right) = \partial\left(-\frac{\rho\bar{\partial}\bar{\rho}}{1+|\rho|^2}\right) + \bar{\partial}\left(\frac{\rho\partial\bar{\rho}}{1+|\rho|^2}\right) = \frac{\bar{\partial}\rho\partial\bar{\rho} - \partial\rho\bar{\partial}\bar{\rho}}{\left(1+|\rho|^2\right)^2}. \tag{87}$$

The integrability condition can be expressed in the form of a zero curvature condition

$$\partial\left(\frac{\partial\bar{\partial}\rho}{\partial\rho} - 2\frac{\bar{\rho}\bar{\partial}\rho}{1+|\rho|^2}\right) - \bar{\partial}\left(\frac{\partial\bar{\partial}\bar{\rho}}{\bar{\partial}\bar{\rho}} - 2\frac{\rho\partial\bar{\rho}}{1+|\rho|^2}\right) = 0. \tag{88}$$

It is clear that provided we have

$$B(\rho,\bar{\rho}) = \partial\bar{\partial}\rho - \frac{2\bar{\rho}}{1+|\rho|^2}\partial\rho\bar{\partial}\rho = 0, \tag{89}$$

the condition is satisfied automatically. This may be recognized as the equation describing the nonlinear sigma model. As well it is the equation which is satisfied by the Gauss map of a constant mean curvature surface which is harmonic.

It is well known that for a given Gauss map ρ such that $\partial\rho = 0$, there is a one parameter family of surfaces called the associated family which is obtained through the transformation

$$\psi_1 \to q^{1/2}\psi_1, \qquad \psi_2 \to \bar{q}^{1/2}\psi_2. \tag{90}$$

This keeps p and ρ invariant if q is a complex constant of modulus one. If $m \neq 0$, it is not possible since m would not stay real. In the latter case, the only allowed values are $q = e^{in\pi}$. To construct the surface, take α and replace the phase factor by $q^{1/2}$, so $p = 1$, and we obtain

$$\psi_1 = \frac{q^{1/2}}{\sqrt{1+|\rho|^2}}, \qquad \psi_2 = -\frac{q^{1/2}\rho}{\sqrt{1+|\rho|^2}}. \tag{91}$$

Substituting (91) into the inducing formulae (30), the Weierstrass representation for $q = 1$ can be observed.

Finally, using (36) and recalling that

$$\partial H = -p^{-2}\bar{\partial}(pS), \qquad \psi_1^2 \partial\bar{\rho} = \frac{i}{2} pS \tag{92}$$

the second equation in (92) is differentiated with respect to $\bar{\partial}$ to obtain,

$$im\psi_1\psi_2\partial\bar{\rho} + \psi_1^2\bar{\partial}\partial\bar{\rho} = \frac{i}{2}\bar{\partial}(pS). \tag{93}$$

Taking the conjugate of the first expression in (72) then solving for ψ_1^2 and substituting into (93) we have

$$-\left(2\frac{\psi_1\psi_2}{p}\partial\bar{\rho} + \frac{\partial\bar{\partial}\rho}{\bar{\partial}\bar{\rho}}\right) = -m^{-1}p^{-1}\bar{\partial}(pS). \tag{94}$$

Using (70) and the relation $|\psi_1|^2/p = 1/\left(1+|\rho|^2\right)$, we obtain the desired result

$$-H^{-1}\partial H = \frac{\partial\bar{\partial}\rho}{\bar{\partial}\bar{\rho}} - 2\frac{\rho\partial\bar{\rho}}{1+|\rho|^2}. \tag{95}$$

Differentiating $J = pS$ with respect to $\bar{\partial}$ then multiplying by $(pS)^{-1}$, we obtain

$$(pS)^{-1}\bar{\partial}(pS) = m\left(2\frac{\psi_1\psi_2}{pS}\partial\bar{\rho} + \frac{1}{S}\frac{\bar{\partial}\partial\bar{\rho}}{\bar{\partial}\bar{\rho}}\partial\bar{\rho}\right) = m\left(\frac{\partial\bar{\rho}}{\bar{\partial}\bar{\rho}}\right)\left(2\frac{\psi_1\psi_2}{pS}\bar{\partial}\bar{\rho} + \frac{1}{S}\frac{\bar{\partial}\partial\bar{\rho}}{\partial\bar{\rho}}\right)$$

$$= \frac{m}{S}\frac{\partial\bar{\rho}}{\bar{\partial}\bar{\rho}}\left(\frac{\bar{\partial}\partial\bar{\rho}}{\partial\bar{\rho}} - 2\frac{\rho\partial\bar{\rho}}{1+|\rho|^2}\right) = \frac{\bar{\partial}\partial\bar{\rho}}{\partial\bar{\rho}} - 2\rho\frac{\partial\bar{\rho}}{1+|\rho|^2}.$$

To obtain this, the first two derivatives in (73) have been used to write $\partial\bar{\rho}/\bar{\partial}\bar{\rho} = S/m$. Summarizing these calculations, the following relations have been proved:

$$-H^{-1}\partial H = \frac{\partial\bar{\partial}\rho}{\bar{\partial}\bar{\rho}} - 2\frac{\rho\partial\bar{\rho}}{1+|\rho|^2} = \frac{\overline{B}}{\bar{\partial}\bar{\rho}}, \qquad J^{-1}\partial J = \frac{\partial\bar{\partial}\bar{\rho}}{\partial\bar{\rho}} - 2\rho\frac{\partial\bar{\rho}}{1+|\rho|^2} = \frac{\overline{B}}{\partial\bar{\rho}}. \tag{96}$$

Thus, for the parameters that are proportional to a power of p, the logarithmic derivatives can still be computed. For a constant mean curvature surface $\partial H = 0$ and so $B(\rho,\bar{\rho}) = 0$ hence

$$\bar{\partial}J = \partial\bar{J} = 0, \tag{97}$$

and the current is conserved, of J is a holomorphic function.

5. Summary and conclusions

It should be said that this work has deep implications for the study of manifolds and their relationship with integrable systems in general [21–24]. It would be worth illustrating this more clearly as a way to conclude. As a particular example, consider the case of a spherical surface for which $S = 0$ so that

$$K = H^2 = \frac{m^2}{p}, \tag{98}$$

where K is now a constant and the Gauss equation simplifies to

$$\partial\bar{\partial}u + m^2 = 0. \tag{99}$$

If we choose $K = 1$, this implies that $m = p$; hence $m = e^u$ and (99) is then the nonlinear Liouville equation

$$\partial\bar{\partial}u + e^{2u} = 0$$

is obtained in terms of the only remaining variable u. This procedure has resulted in a nonlinear equation with a link to surfaces. Since $p^{-1}m = 1$, the Codazzi-Mainardi equation is trivially satisfied.

Due to the spinor representation of the Maurer-Cartan form, from which Z_1 and Z_2 are deduced, for any nonsingular matrix τ, there is a gauge transformation given by [19]

$$\begin{aligned} &\Phi \to \tau\Phi, \\ &Z_1 \to \tau Z_1 \tau^{-1} + \partial\tau \cdot \tau^{-1}, \\ &Z_2 \to \tau Z_2 \tau^{-1} + \bar{\partial}\tau \cdot \tau^{-1}, \end{aligned} \tag{100}$$

for which the nonlinear zero curvature equation still holds. For example, suppose we take

$$\tau = \begin{pmatrix} \bar{\lambda}^{1/2} & 0 \\ 0 & \lambda^{1/2} \end{pmatrix} \begin{pmatrix} e^{u/2} & 0 \\ 0 & e^{-u/2} \end{pmatrix} e^{-u/2}. \tag{101}$$

In (101), λ can be thought of as a complex spectral parameter that satisfies $|\lambda|^2 = 1$. Starting with (8), we find that

$$\Phi'_\lambda = \tau\Phi = \begin{pmatrix} \bar{\lambda}^{1/2}\psi_1 & -\bar{\lambda}^{1/2}\bar{\psi}_2 \\ \lambda^{1/2}e^{-u}\psi_2 & \lambda^{1/2}e^{-u}\bar{\psi}_1 \end{pmatrix}, \qquad \det \Phi'_\lambda = 1. \tag{102}$$

It is straightforward to calculate that

$$\tau Z_1 \tau^{-1} = \frac{1}{2}\begin{pmatrix} 2\partial u & 0 \\ i\lambda & 0 \end{pmatrix}, \quad \partial\tau = \begin{pmatrix} 0 & 0 \\ 0 & -\lambda^{1/2}e^{-u}\partial u \end{pmatrix}, \quad \partial\tau\cdot\tau^{-1} = \begin{pmatrix} 0 & 0 \\ 0 & -\partial u \end{pmatrix},$$

Therefore, we get

$$\tau Z_1 \tau^{-1} + \partial\tau\tau^{-1} = \frac{1}{2}\begin{pmatrix} 2\partial u & 0 \\ i\lambda & -2\partial u \end{pmatrix},$$

and proceeding in a similar fashion, one finds

$$\tau Z_2 \tau^{-1} + \bar{\partial}\tau\tau^{-1} = \frac{1}{2}\begin{pmatrix} 0 & i\bar{\lambda}e^{2u} \\ 0 & 0 \end{pmatrix}.$$

The linear system for the case in which $S = 0$ and $m = p$ is given by

$$\partial\Phi'_\lambda = \frac{1}{2}\begin{pmatrix} 2\partial u & 0 \\ i\lambda & -2\partial u \end{pmatrix}\Phi'_\lambda, \qquad \bar{\partial}\Phi'_\lambda = \frac{1}{2}\begin{pmatrix} 0 & i\bar{\lambda}e^{2u} \\ 0 & 0 \end{pmatrix}\Phi'_\lambda, \tag{103}$$

where Φ'_λ is given by (102). Other choices for the gauge function τ will lead to other systems: for example, taking

$$\tau = \frac{1}{\sqrt{2}}\begin{pmatrix} 1 & -1 \\ 1 & 0 \end{pmatrix}\begin{pmatrix} \bar{\lambda}^{1/2} & 0 \\ 0 & \lambda^{1/2}e^{-u} \end{pmatrix} = \frac{1}{\sqrt{2}}\begin{pmatrix} \bar{\lambda}^{1/2} & -\lambda^{1/2}e^{-u} \\ \bar{\lambda}^{1/2} & \lambda^{1/2}e^{-u} \end{pmatrix} \tag{104}$$

an AKNS type system is obtained

$$\partial\tilde{\Phi} = \frac{1}{4}\begin{pmatrix} -i\lambda & 4\partial u - i\lambda \\ 4\partial u + i\lambda & i\lambda \end{pmatrix}\tilde{\Phi}_\lambda, \qquad \bar{\partial}\tilde{\Phi} = i\frac{\bar{\lambda}}{4}\begin{pmatrix} -e^{2u} & e^{2u} \\ -e^{2u} & e^{2u} \end{pmatrix}\tilde{\Phi}_\lambda. \tag{105}$$

Therefore, hierarchies may be generated and this linear system which is derived from the Dirac equation and used to create surfaces provides the link between nonlinear evolution equations and geometry.

Author details

Paul Bracken

Address all correspondence to: paul.bracken@utrgv.edu

Department of Mathematics, University of Texas, Edinburg, TX, USA

References

[1] Weierstrass K. Fortsetzung der Untersuchung über die Minimalflächer. In: Mathematische Werke. Vol. 3. Berlin: Verlagsbuchhandlung Hillesheim; 1866. pp. 219-248

[2] Konoplechenko BG, Taimanov I. Constant mean curvature surfaces via an integrable dynamical system. Journal of Physics A: Mathematical and General. 1996;**29**:1261-1265

[3] Konopelchenko BG. Induced surfaces and their integrable dynamics. Studies in Applied Mathematics. 1996;**96**:9-51

[4] Carrol R, Konopelchenko BG. Generalized Weierstrass-Enneper systems inducing conformal immersions and gravity. International Journal of Modern Physics A: Particles and Fields; Gravitation; Cosmology; Nuclear Physics. 1996;**11**:1183-1216

[5] Konopelchenko BG, Landolfi G. Generalized Weierstrass representation for surfaces in multi-dimensional Riemann spaces. Journal of Geometry and Physics. 1999;**29**:319-333

[6] Bracken P, Grundland AM. Solutions of the generalized Weierstrass representation in four-dimensional Euclidean space. Journal of Nonlinear Mathematical Physics. 2002;**9**: 357-381

[7] Nelson D, Piran T, Weinberg S. Statistical Mechanics of Membranes and Surfaces. Singapore: World Scientific; 1992

[8] Gross DG, Pope GN, Weinberg S. Two-Dimensional Quantum Gravity and Random Surfaces. Singapore: World Scientific; 1992

[9] Konoplechenko BG, Landolfi G. On classical string configurations. Modern Physics Letters A. 1997;**12**:3161-3168

[10] Viswanathan K, Parthasarathy R. Generalized Gauss map and the geometry of strings. Ann. Phys. 1991;**206**:237-254

[11] Bracken P. A surface model for classical strings in Minkowski space. Physics Letters B. 2002;**541**:166-170

[12] Ablowitz MJ, Kaup DJ, Newell AC, Segur H. The inverse scattering transform-Fourier analysis for nonlinear problems. Studies in Applied Mathematics. 1974;**53**:249-315

[13] Hoffman DA, Osserman R. The Gauss map of surfaces in $\mathbb{R}^3$ and $\mathbb{R}^4$. Proceedings of the London Mathematical Society. 1985;**50**:27-56

[14] Bracken P. In: Lang C, editor. The Generalized Weierstrass System Inducing Surfaces in Euclidean Three Space and Higher Dimensional Spaces: Partial Differential Equations. Hauppaugo, NY: Nova Science Publishers; 2011

[15] Bracken P, Grundland AM, Martina L. The Weierstrass-Enneper system for constant mean curvature surfaces and the completely integrable Sigma model. Journal of Mathematical Physics. 1999;**40**:3379-3403

[16] Bracken P, Grundland AM. Symmetry properties and explicit solutions of the generalized Weierstrass system. Journal of Mathematical Physics. 2001;**42**:1250-1282

[17] Ody M, Ryder L. Time-independent solutions to the two-dimensional nonlinear $O(3)$ Sigma model and surfaces of constant mean curvature. International Journal of Modern Physics A: Particles and Fields; Gravitation; Cosmology; Nuclear Physics. 1995;**10**:337-364

[18] Kenmotsu K. Weierstrass formula for surfaces of prescribed mean curvature. Mathematische Annalen. 1979;**245**:89-99

[19] Bracken P. A geometric formulation of lax integrability for nonlinear equations. Tensor. 2015;**76**(2):145-156

[20] Bracken P. The generalized Weierstrass system for nonconstant mean curvature surfaces and the nonlinear Sigma model. Acta Applicandae Mathematicae. 2006;**92**:63-76

[21] Shipman B, Shipman P, Packard D. Generalized Weierstrass-Enneper representations of Euclidean spacelike and timelike surfaces: A unified lie-algebraic formulation. Journal of Geometry. 2017;**108**:545-563

[22] Ganchev G, Kanchev K. Canonical Weierstrass representations for minimal space-like surfaces in $\mathbb{R}_1^4$. 2017. arXiv: 1612.05504v2

[23] Cintra A, Mercuri F, Onnis I. Minimal surfaces in Lorentzian Heisenberg group and Damek-Ricci spaces via the Weierstrass representation. Journal of Geometry and Physics. 2017;**121**:396-412

[24] Fomenko AT, Tuzhlin AA. Elements of the geometry and topology of minimal surfaces in three-dimensional space. In: Mathematical Monographs. Vol. 93. Providence, RI: AMS; 2005

Chapter 2

On Conformal Anti-Invariant Submersions Whose Total Manifolds Are Locally Product Riemannian

Mehmet Akif Akyol

Additional information is available at the end of the chapter

http://dx.doi.org/10.5772/intechopen.80337

Abstract

The aim of this chapter is to study conformal anti-invariant submersions from almost product Riemannian manifolds onto Riemannian manifolds as a generalization of anti-invariant Riemannian submersion which was introduced by B. Sahin. We investigate the integrability of the distributions which arise from the definition of the new submersions and the geometry of foliations. Moreover, we find necessary and sufficient conditions for this submersion to be totally geodesic and in order to guarantee the new submersion, we mention some examples of such submersions.

Keywords: conformal submersion, almost product Riemannian manifold, vertical distribution, conformal anti-invariant submersion
2010 Mathematics Subject Classification: primary 53C15; secondary 53C40

1. Introduction

Immersions and submersions, which are special tools in differential geometry, also play a fundamental role in Riemannian geometry, especially when the involved manifolds carry an additional structure (such as contact, Hermitian and product structure). In particular, Riemannian submersions (which we always assume to have connected fibers) are fundamentally important in several areas of Riemannian geometry. For instance, it is a classical and important problem in Riemannian geometry to construct Riemannian manifolds with positive or non-negative sectional curvature. Riemannian submersions between Riemannian manifolds are important geometric structures. Riemannian submersions between Riemannian manifolds were studied by O'Neill [1] and Gray [2]. In [3], the Riemannian submersions were considered between almost Hermitian manifolds by Watson under the name of almost Hermitian submersions.

In this case, the Riemannian submersion is also an almost complex mapping and consequently the vertical and horizontal distributions are invariant with respect to the almost complex structure of the total manifold of the submersion. The study of anti-invariant Riemannian submersions from almost Hermitian manifolds was initiated by Şahin [4]. In this case, the fibers are anti-invariant with respect to the almost complex structure of the total manifold. This notion extended to different total spaces see: [5–14].

On the other hand, as a generalization of Riemannian submersion, horizontally conformal submersions are defined as follows [15]: Suppose that (M, g_M) and (B, g_B) are Riemannian manifolds and $\pi : M \to B$ is a smooth submersion, then π is called a horizontally conformal submersion, if there is a positive function λ such that

$$\lambda^2 g_M(X, Y) = g_B(\pi_* X, \pi_* Y)$$

for every $X, Y \in \Gamma\left((ker\pi_*)^{\perp}\right)$. It is obvious that every Riemannian submersion is a particular horizontally conformal submersion with $\lambda = 1$. We note that horizontally conformal submersions are special horizontally conformal maps which were introduced independently by Fuglede [16] and Ishihara [17]. We also note that a horizontally conformal submersion $\pi : M \to B$ is said to be horizontally homothetic if the gradient of its dilation λ is vertical, i.e.,

$$\mathcal{H}(grad\lambda) = 0 \tag{1}$$

at $p \in M$, where $\mathcal{H}$ is the projection on the horizontal space $(ker\pi_*)^{\perp}$. For conformal submersion, see: [15, 18, 19].

One can see that Riemannian submersions are very special maps comparing with conformal submersions. Although conformal maps do not preserve distance between points contrary to isometries, they preserve angles between vector fields. This property enables one to transfer certain properties of a manifold to another manifold by deforming such properties.

Recently, we introduced conformal anti-invariant submersions [20] and conformal semi-invariant submersions [21] from almost Hermitian manifolds, and gave examples and investigated the geometry of such submersions (see also [22, 23]). We showed that the geometry of such submersions is different from their counterpart anti-invariant Riemannian submersions and semi-invariant Riemannian submersions. In the present paper, we define and study conformal anti-invariant submersions from almost product Riemannian manifolds, give examples and investigate the geometry of the total space and the base space for the existence of such submersions.

Our work is structured as follows: Section 2 is focused on basic facts for conformal submersions and almost product Riemannian manifolds. The third section is concerned with definition of conformal anti-invariant submersions, investigating the integrability conditions of the horizontal distribution and the vertical distribution. In Section 4, we study the geometry of leaves of the horizontal distribution and the vertical distribution. In Section 5, we find necessary and sufficient conditions for a conformal anti-invariant submersion to be totally geodesicness. The last section, we give some examples of such submersions.

2. Preliminaries

In this section we recall several notions and results which will be needed throughout the chapter.

Let M be a m-dimensional manifold with a tensor F of a type (1,1) such that

$$F^2 = I, (F \neq I).$$

Then, we say that M is an almost product manifold with almost product structure F. We put

$$P = \frac{1}{2}(I + F), \quad Q = \frac{1}{2}(I - F).$$

Then we get

$$P + Q = I, \quad P^2 = P, \quad Q^2 = Q, \quad PQ = QP = 0, \quad F = P - Q.$$

Thus P and Q define two complementary distributions P and Q. We easily see that the eigenvalues of F are +1 or -1. If an almost product manifold M admits a Riemannian metric g such that

$$g(FX, FY) = g(X, Y) \tag{2}$$

for any vector fields X and Y on M, then M is called an almost product Riemannian manifold, denoted by (M, g, F). Denote the Levi-Civita connection on M with respect to g by ∇. Then, M is called a locally product Riemannian manifold [24] if F is parallel with respect to ∇, i.e.,

$$\nabla_X F = 0, \quad X \in \Gamma(TM). \tag{3}$$

Conformal submersions belong to a wide class of conformal maps that we are going to recall their definition, but we will not study such maps in this paper.

Definition 2.1 ([15]) *Let $\varphi : (M^m, g) \to (N^n, h)$ be a smooth map between Riemannian manifolds, and let $x \in M$. Then φ is called horizontally weakly conformal or semi conformal at x if either*

(i) $d\varphi_x = 0$, *or*

(ii) $d\varphi_x$ *maps horizontal space* $\mathcal{H}_x = \left(ker(d\varphi_x)\right)^{\perp}$ *conformally onto* $T_{\varphi_*}N$, *i.e.,* $d\varphi_x$ *is surjective and there exists a number* $\Lambda(x) \neq 0$ *such that*

$$h\big(d\varphi_x X, d\varphi_x Y\big) = \Lambda(x)g(X, Y) \quad (X, Y \in \mathcal{H}_x). \tag{4}$$

Note that we can write the last equation more sufficiently as

$$(\varphi^* h)_x\big|_{\mathcal{H}_x \times \mathcal{H}_x} = \Lambda(x) g_x\big|_{\mathcal{H}_x \times \mathcal{H}_x}.$$

A point x is of type (i) in Definition if and only if it is a critical point of φ; we shall call a point of type (ii) a *regular point*. At a critical point, $d\varphi_x$ has rank 0; at a regular point, $d\varphi_x$ has rank n and φ is submersion. The number $\Lambda(x)$ is called the *square dilation* (of φ at x); it is necessarily non-negative; its square root $\lambda(x) = \sqrt{\Lambda(x)}$ is called the dilation (of φ at x). The map φ is called *horizontally weakly conformal* or *semi conformal* (on M) if it is horizontally weakly conformal at every point of M. It is clear that if φ has no critical points, then we call it a (*horizontally*) conformal submersion.

Next, we recall the following definition from [18]. Let $\pi : M \to N$ be a submersion. A vector field E on M is said to be projectable if there exists a vector field $\breve{E}$ on N, such that $d\pi(E_x) = \breve{E}_{\pi(x)}$ for all $x \in M$. In this case E and $\breve{E}$ are called $\pi-$ related. A horizontal vector field Y on (M, g) is called basic, if it is projectable. It is well known fact, that is, $\breve{Z}$ is a vector field on N, then there exists a unique basic vector field Z on M, such that Z and $\breve{Z}$ are $\pi-$ related. The vector field Z is called the horizontal lift of $\breve{Z}$.

The fundamental tensors of a submersion were introduced in [1]. They play a similar role to that of the second fundamental form of an immersion. More precisely, O'Neill's tensors T and A defined for vector fields E, G on M by

$$A_E G = \mathcal{V}\nabla^{M_1}_{\mathcal{H}E}\mathcal{H}G + \mathcal{H}\nabla^{M_1}_{\mathcal{H}E}\mathcal{V}G \tag{5}$$

$$T_E G = \mathcal{H}\nabla^{M_1}_{\mathcal{V}E}\mathcal{V}G + \mathcal{V}\nabla^{M_1}_{\mathcal{V}E}\mathcal{H}G \tag{6}$$

where $\mathcal{V}$ and $\mathcal{H}$ are the vertical and horizontal projections (see [25]). On the other hand, from (5) and (6), we have

$$\nabla^{M_1}_V W = T_V W + \hat{\nabla}_V W \tag{7}$$

$$\nabla^{M_1}_V X = \mathcal{H}\nabla^{M_1}_V X + T_V X \tag{8}$$

$$\nabla^{M_1}_X V = A_X V + \mathcal{V}\nabla^{M_1}_X V \tag{9}$$

$$\nabla^{M_1}_X Y = \mathcal{H}\nabla^{M_1}_X Y + A_X Y \tag{10}$$

for $X, Y \in \Gamma\big((ker\pi_*)^\perp\big)$ and $V, W \in \Gamma(ker\pi_*)$, where $\hat{\nabla}_V W = \mathcal{V}\nabla^{M_1}_V W$. If X is basic, then $\mathcal{H}\nabla^{M_1}_V X = A_X V$. It is easily seen that for $x \in M$, $X \in \mathcal{H}_x$ and $\mathcal{V}_x$ the linear operators T_V, $A_X : T_X M \to T_X M$ are skew-symmetric, that is

$$g(T_V E, G) = -g(E, T_V G) \text{ and } g(A_X E, G) = -g(E, A_X G)$$

for all $E, G \in T_x M$. We also see that the restriction of T to the vertical distribution $T|_{V \times V}$ is exactly the second fundamental form of the fibers of π. Since T_V is skew symmetric, we get π which has totally geodesic fibers if and only if $T \equiv 0$. For the special case when π is horizontally conformal we have the following:

Proposition 2.1 ([18]) *Let $\pi : (M^m, g) \to (N^n, h)$ be a horizontally conformal submersion with dilation ∇ and X, Y be horizontal vectors, then*

$$A_X Y = \frac{1}{2}\left\{\mathcal{V}[X, Y] - \lambda^2 g(X, Y) grad_{\mathcal{V}}\left(\frac{1}{\lambda^2}\right)\right\}. \tag{11}$$

We see that the skew-symmetric part of $A|_{(ker\pi_*)^\perp \times (ker\pi_*)^\perp}$ measures the obstruction integrability of the horizontal distribution $(ker\pi_*)^\perp$.

Let (M, g_M) and (N, g_N) be Riemannian manifolds and suppose that $\pi : M \to N$ is a smooth map between them. The differential of π_* of π can be viewed a section of the bundle $Hom(TM, \pi^{-1}TN) \to M$, where $\pi^{-1}TN$ is the pullback bundle which has fibers $(\pi^{-1}TN)_p = T_{\pi(p)}N$, $p \in M$. $Hom(TM, \pi^{-1}TN)$ has a connection ∇ induced from the Levi-Civita connection ∇^M and the pullback connection. Then the second fundamental form of π is given by

$$\nabla\pi_* : \Gamma(TM) \times \Gamma(TM) \to \Gamma(TN)$$

defined by

$$(\nabla\pi_*)(X, Y) = \nabla^\pi_X \pi_*(Y) - \pi_*(\nabla^M_X Y) \tag{12}$$

for $X, Y \in \Gamma(TM)$, where ∇^π is the pullback connection. It is known that the second fundamental form is symmetric.

Lemma 2.1. [26] *Let (M, g_M) and (N, g_N) be Riemannian manifolds and suppose that $\varphi : M \to N$ is a smooth map between them. Then we have*

$$\nabla^\varphi_X \varphi_*(Y) - \nabla^\varphi_Y \varphi_*(X) - \varphi_*([X, Y]) = 0 \tag{13}$$

for $X, Y \in \Gamma(TM)$.

Finally, we recall the following lemma from [15].

Lemma 2.2. *Suppose that $\pi : M \to N$ is a horizontally conformal submersion. Then, for any horizontal vector fields X, Y and vertical fields V, W we have.*

(i) $(\nabla\pi_*)(X,Y) = X(\ln\lambda)\pi_*Y + Y(\ln\lambda)\pi_*X - g_M(X,Y)\pi_*(grad\ln\lambda)$;

(ii) $(\nabla\pi_*)(V,W) = -\pi_*(T_VW)$;

(iii) $(\nabla\pi_*)(X,V) = -\pi_*(\nabla_X^M V) = -\pi_*(A_XV)$.

3. Conformal anti-invariant submersions from almost product Riemannian manifolds

In this section, we define conformal anti-invariant submersions from an almost product Riemannian manifold onto a Riemannian manifold, investigating the geometry of distributions $(ker\pi_*)$ and $(ker\pi_*)^\perp$ and obtain the integrability conditions for the distribution $(ker\pi_*)^\perp$ for such submersions.

Definition 3.1. *Let (M_1,g_1,F) be an almost product Riemannian manifold and (M_2,g_2) be a Riemannian manifold. A horizontally conformal submersion $\pi: M_1 \to M_2$ with dilation λ is called conformal anti-invariant submersion if the distribution $ker\pi_*$ is anti-invariant with respect to F, i.e., $F(ker\pi_*) \subseteq (ker\pi_*)^\perp$.*

Let $\pi:(M_1,g_1,F)\to(M_2,g_2)$ is a conformal anti-invariant submersion from an almost product Riemannian manifold (M_1,g_1,F) to a Riemannian manifold (M_2,g_2). First of all, from Definition 3.1, we have $F(ker\pi_*)^\perp \cap ker\pi_* \neq 0$. We denote the complementary orthogonal distribution to $F(ker\pi_*)$ in $(ker\pi_*)^\perp$ by μ. Then we have

$$(ker\pi_*)^\perp = F(ker\pi_*) \oplus \mu. \tag{14}$$

Proposition 3.1. *Let (M_1,g_1,F) be an almost product Riemannian manifold and (M_2,g_2) be a Riemannian manifold. Then μ is invariant with respect to F.*

Proof. For $Z\in\Gamma(\mu)$ and $V\in\Gamma(ker\pi_*)$, by using (2), we have $g_1(FZ,FV)=0$, which show that FZ is orthogonal to $Fker\pi_*$. On the other hand, since FV and Z are orthogonal we get $g_1(FV,Z) = g_1(V,FZ) = 0$ which shows that FZ is orthogonal to $ker\pi_*$. This completes proof. □

For $Z\in\Gamma\left((ker\pi_*)^\perp\right)$, we have

$$FZ = \mathcal{B}Z + \mathcal{C}Z, \tag{15}$$

where $\mathcal{B}Z\in\Gamma(ker\pi_*)$ and $\mathcal{C}Z\in\Gamma(\mu)$. On the other hand, since $\pi_*\left((ker\pi_*)^\perp\right) = TM_2$ and π is a conformal submersion, using (15) we derive $\frac{1}{\lambda^2}g_2(\pi_*FV,\pi_*\mathcal{C}Z)=0$ for any $Z\in\Gamma\left((ker\pi_*)^\perp\right)$ and $V\in\Gamma(ker\pi_*)$, which implies that

$$TM_2 = \pi_*(Fker\pi_*) \oplus \pi_*(\mu). \tag{16}$$

Lemma 3.1. *Let π be a conformal anti-invariant submersion from a locally product Riemannian manifold (M_1, g_1, F) onto a Riemannian manifold (M_2, g_2). Then we have*

$$g_1(\mathcal{C}W, FV) = 0 \tag{17}$$

and

$$g_1(\nabla^{M_1}_Z \mathcal{C}W, FV) = -g_1(\mathcal{C}W, FA_Z V) \tag{18}$$

for $Z, W \in \Gamma\left((ker\pi_*)^\perp\right)$ and $V \in \Gamma(ker\pi_*)$.

Proof. For $W \in \Gamma\left((ker\pi_*)^\perp\right)$ and $V \in \Gamma(ker\pi_*)$, using (2) we have

$$g_1(\mathcal{C}W, FV) = g_1(FW - \mathcal{B}W, FV) = g_1(FW, FV)$$

due to $\mathcal{B}W \in \Gamma(ker\pi_*)$ and $FV \in \Gamma\left((ker\pi_*)^\perp\right)$. Hence $g_1(FW, FV) = g_1(W, V) = 0$ which is (17). Since M_1 is a locally product Riemannian manifold, differentiating (3.4) with respect to Z, we get

$$g_1(\nabla^{M_1}_Z \mathcal{C}W, FV) = g_1(\mathcal{C}W, F\nabla^{M_1}_Z V)$$

for $Z, W \in \Gamma\left((ker\pi_*)^\perp\right)$ and $V \in \Gamma(ker\pi_*)$. Then using (9) we have

$$g_1(\nabla^{M_1}_Z \mathcal{C}W, FV) = -g_1(\mathcal{C}W, FA_Z V) - g_1(\mathcal{C}W, F\mathcal{V}\nabla^{M_1}_Z V).$$

Since $F\mathcal{V}\nabla^{M_1}_Z V \in \Gamma(Fker\pi_*)$, we obtain (18). □

We now study the integrability of the distribution $(\ker\pi_*)^\perp$ and then we investigate the geometry of the leaves of $\ker\pi_*$ and $(\ker\pi_*)^\perp$. We note that it is known that the distribution $\ker\pi_*$ is integrable.

Theorem 3.1. *Let $\pi : (M_1, g_1, F) \to (M_2, g_2)$ is a conformal anti-invariant submersion from an almost product Riemannian manifold (M_1, g_1, F) to a Riemannian manifold (M_2, g_2). Then the following assertions are equivalent to each other;*

(a) *$(ker\pi_*)^\perp$ is integrable,*

(b) $$\frac{1}{\lambda^2} g_2(\nabla^\pi_W \pi_*\mathcal{C}Z - \nabla^\pi_Z \pi_*\mathcal{C}W, \pi_*FV) = g_1(A_Z\mathcal{B}W - A_W\mathcal{B}Z - \mathcal{C}W(\ln\lambda)Z + \mathcal{C}Z(\ln\lambda)W, FV) \tag{19}$$

for any $Z, W \in \Gamma\left((ker\pi_*)^\perp\right)$ and $V \in \Gamma(ker\pi_*)$.

Proof. For $W \in \Gamma\left((ker\pi_*)^\perp\right)$ and $V \in \Gamma(ker\pi_*)$, we see from Definition 3.1, $FV \in \Gamma\left((ker\pi_*)^\perp\right)$ and $FW \in \Gamma(ker\pi_* \oplus \mu)$. Thus using (2) and (3), for $Z \in \Gamma\left((ker\pi_*)^\perp\right)$ we obtain

$$g_1([Z,W],V) = g_1(\nabla_Z^{M_1} FW, FV) - g_1(\nabla_W^{M_1} FZ, FV).$$

Further, from (15) we get

$$g_1([Z,W],V) = g_1(\nabla_Z^{M_1} \mathcal{B}W, FV) + g_1(\nabla_Z^{M_1} \mathcal{C}W, FV) - g_1(\nabla_W^{M_1} \mathcal{B}Z, FV) - g_1(\nabla_W^{M_1} \mathcal{C}Z, FV).$$

Using (9), (10) and if we take into account π is a conformal submersion, we arrive at

$$g_1([Z,W],V) = g_1(A_Z\mathcal{B}W - A_W\mathcal{B}Z, FV) + \frac{1}{\lambda^2} g_2(\pi_*(\nabla_Z^{M_1}\mathcal{C}W), \pi_*FV) - \frac{1}{\lambda^2} g_2(\pi_*(\nabla_W^{M_1}\mathcal{C}Z), \pi_*FV).$$

Thus, from (12) and Lemma 2.2 we derive

$$\begin{aligned} g_1([Z,W],V) = {} & g_1(A_Z\mathcal{B}W - A_W\mathcal{B}Z, FV) - g_1(\mathcal{H}grad \ln\lambda, Z) g_1(\mathcal{C}W, FV) \\ & - g_1(\mathcal{H}grad \ln\lambda, \mathcal{C}W) g_1(Z, FV) + g_1(Z, \mathcal{C}W) g_1(\mathcal{H}grad \ln\lambda, FV) \\ & + \frac{1}{\lambda^2} g_2(\nabla_Z^{\pi}\pi_*\mathcal{C}W, \pi_*FV) + g_1(\mathcal{H}grad \ln\lambda, W) g_1(\mathcal{C}Z, FV) \\ & + g_1(\mathcal{H}grad \ln\lambda, \mathcal{C}Z) g_1(W, FV) - g_1(W, \mathcal{C}Z) g_1(\mathcal{H}grad \ln\lambda, FV) \\ & - \frac{1}{\lambda^2} g_2(\nabla_W^{\pi}\pi_*\mathcal{C}Z, \pi_*FV). \end{aligned}$$

Moreover, using (17), we obtain

$$\begin{aligned} g_1([Z,W],V) = {} & g_1(A_Z\mathcal{B}W - A_W\mathcal{B}Z - \mathcal{C}W(\ln\lambda)Z + \mathcal{C}Z(\ln\lambda)W, FV) \\ & - \frac{1}{\lambda^2} g_2(\nabla_W^{\pi}\pi_*\mathcal{C}Z - \nabla_Z^{\pi}\pi_*\mathcal{C}W, \pi_*FV) \end{aligned}$$

which proves $(a) \Leftrightarrow (b)$. □

From Theorem 3.1, we deduce the following characterization.

Theorem 3.2. *Let π be a conformal anti-invariant submersion from a locally product Riemannian manifold (M_1, g_1, F) onto a Riemannian manifold (M_2, g_2). Then any two conditions below imply the three;*

i. $(ker\pi_*)^\perp$ *is integrable.*

ii. λ *is a constant on* $\Gamma(\mu)$.

iii. $g_2(\nabla_W^{\pi}\pi_*\mathcal{C}Z - \nabla_Z^{\pi}\pi_*\mathcal{C}W, \pi_*FV) = \lambda^2 g_1(A_Z\mathcal{B}W - A_W\mathcal{B}Z, FV)$

for $Z, W \in \Gamma\left((ker\pi_*)^\perp\right)$ *and* $V \in \Gamma(ker\pi_*)$.

Proof. From Theorem 3.1, we have

$$g_1([Z,W],V) = g_1(A_Z\mathcal{B}W - A_W\mathcal{B}Z - \mathcal{C}W(\ln\lambda)Z + \mathcal{C}Z(\ln\lambda)W, FV) - \frac{1}{\lambda^2}g_2(\nabla^\pi_W\pi_*\mathcal{C}Z - \nabla^\pi_Z\pi_*\mathcal{C}W, \pi_*FV).$$

for $Z, W \in \Gamma\left((ker\pi_*)^\perp\right)$ and $V \in \Gamma(ker\pi_*)$. Now, if we have (i) and (iii), then we arrive at

$$-g_1(\mathcal{H}grad\ln\lambda, \mathcal{C}W)g_1(Z, FV) + g_1(\mathcal{H}grad\ln\lambda, \mathcal{C}Z)g_1(W, FV) = 0. \tag{20}$$

Now, taking $W = FV$ in (20) for $V \in \Gamma(ker\pi_*)$, using (17), we get

$$-g_1(\mathcal{H}grad\ln\lambda, \mathcal{C}(FV))g_1(Z, FV) + g_1(\mathcal{H}grad\ln\lambda, \mathcal{C}Z)g_1(FV, FV) = 0.$$

Hence λ is a constant on $\Gamma(\mu)$. Similarly, one can obtain the other assertions. □

We say that a conformal anti-invariant submersion is a conformal Lagrangian submersion if $F(ker\pi_*) = (ker\pi_*)^\perp$. From Theorem 3.1, we have the following result.

Corollary 3.1. *Let* π *be a conformal Lagrangian submersion from a locally product Riemannian manifold* (M_1, g_1, F) *onto a Riemannian manifold* (M_2, g_2). *Then the following assertions are equivalent to each other:*

i. $(ker\pi_*)^\perp$ *is integrable*

ii. $A_Z FW = A_W FZ$

iii. $(\nabla\pi_*)(Z, FW) = (\nabla\pi_*)(W, FZ)$

for $Z, W \in \Gamma\left((ker\pi_*)^\perp\right)$.

Proof. For $Z, W \in \Gamma\left((ker\pi_*)^\perp\right)$ and $V \in \Gamma(ker\pi_*)$, we see from Definition 3.1, $FV \in \Gamma\left((ker\pi_*)^\perp\right)$ and $FW \in \Gamma(ker\pi_*)$. From Theorem 3.1 we have

$$g_1([Z,W],V) = g_1(A_Z\mathcal{B}W - A_W\mathcal{B}Z - \mathcal{C}W(\ln\lambda)Z + \mathcal{C}Z(\ln\lambda)W, FV) - \frac{1}{\lambda^2}g_2(\nabla^\pi_W\pi_*\mathcal{C}Z - \nabla^\pi_Z\pi_*\mathcal{C}W, \pi_*FV).$$

Since π is a conformal Lagrangian submersion, we derive

$$g_1([Z,W],V) = g_1(A_Z\mathcal{B}W - A_W\mathcal{B}Z, FV)$$

which shows $(i) \Leftrightarrow (ii)$. On the other hand, using Definition 3.1 and (9) we arrive at

$$\begin{aligned} g_1(A_Z\mathcal{B}W,FV)-g_1(A_W\mathcal{B}Z,FV) &= \frac{1}{\lambda^2}g_2(\pi_*A_Z\mathcal{B}W,\pi_*FV)-\frac{1}{\lambda^2}g_2(\pi_*A_W\mathcal{B}Z,\pi_*FV) \\ &= \frac{1}{\lambda^2}g_2\left(\pi_*\left(\nabla^{M_1}_Z\mathcal{B}W\right),\pi_*FV\right)-\frac{1}{\lambda^2}g_2\left(\pi_*\left(\nabla^{M_1}_W\mathcal{B}Z\right),\pi_*FV\right). \end{aligned}$$

Now, using (12) we obtain

$$\frac{1}{\lambda^2}\left\{g_2\left(\pi_*\left(\nabla^{M_1}_Z\mathcal{B}W\right),\pi_*FV\right)-g_2\left(\pi_*\left(\nabla^{M_1}_W\mathcal{B}Z\right),\pi_*FV\right)\right\}$$

$$=\frac{1}{\lambda^2}g_2\left(-(\nabla F_*)(Z,\mathcal{B}W)+\nabla^{\pi}_Z\pi_*\mathcal{B}W,\pi_*FV\right)-\frac{1}{\lambda^2}g_2\left(-(\nabla F_*)(W,\mathcal{B}Z)+\nabla^{\pi}_W\pi_*\mathcal{B}Z,\pi_*FV\right).$$

Since $BZ, BW\in\Gamma(ker\pi_*)$, we derive

$$g_1(A_Z\mathcal{B}W,FV)-g_1(A_W\mathcal{B}Z,FV)=\frac{1}{\lambda^2}g_2((\nabla F_*)(W,\mathcal{B}Z)-(\nabla F_*)(Z,\mathcal{B}W),\pi_*FV)$$

which tells that $(ii)\Leftrightarrow(iii)$. □

4. Totally geodesic foliations

In this section, we shall investigate the geometry of leaves of $(ker\pi_*)$ and $(ker\pi_*)^{\perp}$. For the geometry of leaves of the horizontal distribution $(ker\pi_*)^{\perp}$, we have the following theorem.

Theorem 4.1. *Let $\pi:(M_1,g_1,F)\to(M_2,g_2)$ is a conformal anti-invariant submersion from an almost product Riemannian manifold (M_1,g_1,F) to a Riemannian manifold (M_2,g_2). Then the following assertions are equivalent to each other;*

i. $(ker\pi_*)^{\perp}$ defines a totally geodesic foliation on M_1.

ii. $-\frac{1}{\lambda^2}g_2(\nabla^{\pi}_Z\pi_*CW,\pi_*FV) \;=g_1(A_ZBW-CW(\ln\lambda)Z+g_1(Z,CW)\ln\lambda,FV)$

for $Z,W\in\Gamma\left((ker\pi_*)^{\perp}\right)$ and $V\in\Gamma(ker\pi_*)$.

Proof. For $Z,W\in\Gamma\left((ker\pi_*)^{\perp}\right)$ and $V\in\Gamma(ker\pi_*)$, by using (3), (9), (10), (14) and (15) we have

$$g_1\left(\nabla^{M_1}_ZW,V\right)=g_1(A_ZBW,FV)+g_1\left(\nabla^{M_1}_ZCW,FV\right).$$

Since π is a conformal submersion, using (12) and Lemma 2.2 we arrive at

$$
\begin{aligned}
g_1\left(\nabla_Z^{M_1} W, V\right) = {} & g_1(A_Z BW, FV) - \frac{1}{\lambda^2} g_1(\mathcal{H} grad \ln \lambda, Z) g_2(\pi_* CW, \pi_* FV) \\
& - \frac{1}{\lambda^2} g_1(\mathcal{H} grad \ln \lambda, CW) g_2(\pi_* Z, \pi_* FV) \\
& + \frac{1}{\lambda^2} g_1(Z, CW) g_2(\pi_*(\mathcal{H} grad \ln \lambda), \pi_* FV) \\
& + \frac{1}{\lambda^2} g_2\left(\nabla_Z^{\pi} \pi_* CW, \pi_* FV\right).
\end{aligned}
$$

Moreover, using Definition 3.1 and (17) we obtain

$$
g_1\left(\nabla_Z^{M_1} W, V\right) = g_1\left(A_Z BW - CW(\ln \lambda) Z + g_1(Z, CW) \ln \lambda, FV\right) + \frac{1}{\lambda^2} g_2\left(\nabla_Z^{\pi} \pi_* CW, \pi_* FV\right)
$$

which proves $(i) \Leftrightarrow (ii)$. □

From Theorem 4.1, we also deduce the following characterization.

Theorem 4.2. *Let π be a conformal anti-invariant submersion from a locally product Riemannian manifold (M_1, g_1, F) onto a Riemannian manifold (M_2, g_2). Then any two conditions below imply the three;*

i. $(ker\pi_*)^{\perp}$ *defines a totally geodesic foliation on* M_1.

ii. π *is horizontally homothetic submersion.*

iii. $g_2\left(\nabla_Z^{\pi} \pi_* CW, \pi_* FV\right) = \lambda^2 g_1(A_Z FV, \mathcal{B}W)$

for $Z, W \in \Gamma\left((ker\pi_*)^{\perp}\right)$ and $V \in \Gamma(ker\pi_*)$.

Proof. For $Z, W \in \Gamma\left((ker\pi_*)^{\perp}\right)$ and $V \in \Gamma(ker\pi_*)$, from Theorem 4.1, we have

$$
g_1\left(\nabla_Z^{M_1} W, V\right) = g_1\left(A_Z BW - CW(\ln \lambda) Z + g_1(Z, CW) \ln \lambda, FV\right) + \frac{1}{\lambda^2} g_2\left(\nabla_Z^{\pi} \pi_* CW, \pi_* FV\right).
$$

Now, if we have (i) and (iii), then we obtain

$$
-g_1(\mathcal{H} grad \ln \lambda, CW) g_1(Z, FV) + g_1(\mathcal{H} grad \ln \lambda, FV) g_1(Z, CW) = 0. \tag{21}
$$

Now, taking $Z = CW$) in (4.1) and using (17), we get

$$
g_1(\mathcal{H} grad \ln \lambda, FV) g_M(CW, CW) = 0.
$$

Thus, λ is a constant on $\Gamma(Fker\pi_*)$. On the other hand, taking $Z = FV$ in (25) and using (17) we derive

$$g_1(\mathcal{H}grad\ln\lambda, CW)g_1(FV, FV) = 0.$$

From above equation, λ is a constant on $\Gamma(\mu)$. Similarly, one can obtain the other assertions. □

For conformal Lagrangian submersion, we have the following result.

Corollary 4.1. *Let π be a conformal Lagrangian submersion from a locally product Riemannian manifold (M_1, g_1, F) onto a Riemannian manifold (M_2, g_2). Then the following assertions are equivalent to each other;*

i. $(ker\pi_*)^{\perp}$ *defines a totally geodesic foliation on* M_1.

ii. $A_Z\mathcal{B}W = 0$

iii. $(\nabla\pi_*)(Z, FV) = 0$

for $Z, W \in \Gamma\left((ker\pi_*)^{\perp}\right)$ and $V \in \Gamma(ker\pi_*)$.

Proof. For $Z, W \in \Gamma\left((ker\pi_*)^{\perp}\right)$ and $V \in \Gamma(ker\pi_*)$, from Theorem 4.1, we have

$$g_1(\nabla_Z^{M_1}W, V) = g_1(A_ZBW - CW(\ln\lambda)Z + g_1(Z, CW)\ln\lambda, FV) + \frac{1}{\lambda^2}g_2(\nabla_Z^{\pi}\pi_*CW, \pi_*FV).$$

Since π is a conformal Lagrangian submersion, we derive

$$g_1(\nabla_Z^{M_1}W, V) = g_1(A_ZBW, FV)$$

which shows $(i) \Leftrightarrow (ii)$. On the other hand, using Definition 3.1 and (9) we arrive at

$$g_1(A_ZBW, FV) = \frac{1}{\lambda^2}g_2(\pi_*(A_Z\mathcal{B}W), \pi_*FV) = \frac{1}{\lambda^2}g_2(\pi_*(\nabla_Z^{M_1}\mathcal{B}W), \pi_*FV).$$

Now, using (12) we obtain

$$\begin{aligned}\frac{1}{\lambda^2}g_2(\pi_*(\nabla_Z^{M_1}\mathcal{B}W), \pi_*FV) &= \frac{1}{\lambda^2}g_2(-(\nabla\pi_*)(Z, \mathcal{B}W) + \nabla_Z^{\pi}\pi_*\mathcal{B}W, \pi_*FV)\\ &= -\frac{1}{\lambda^2}g_2((\nabla\pi_*)(Z, \mathcal{B}W), \pi_*FV)\end{aligned}$$

which tells that $(ii) \Leftrightarrow (iii)$. □

For the totally geodesicness of the foliations of the distribution $ker\pi_*$.

Theorem 4.3. *Let $\pi : (M_1, g_1, F) \rightarrow (M_2, g_2)$ is a conformal anti-invariant submersion from an almost product Riemannian manifold (M_1, g_1, F) to a Riemannian manifold (M_2, g_2). Then the following assertions are equivalent to each other;*

i. *$ker\pi_*$ defines a totally geodesic foliation on M_1.*

ii. $-\frac{1}{\lambda^2} g_2(\nabla^{\pi}_{FU}\pi_* FV, \pi_* FCZ) \quad = g_1(T_V FU, \mathcal{B}Z) + g_1(U,V) g_1(\mathcal{H}grad \ln\lambda, FCZ)$

for $V, U \in \Gamma(ker\pi_*)$ and $Z \in \Gamma\left((ker\pi_*)^{\perp}\right)$.

Proof. For $Z \in \Gamma\left((ker\pi_*)^{\perp}\right)$ and $V, U \in \Gamma(ker\pi_*)$, by using (2), (3), (8) and (15) we get

$$g_1(\nabla^{M_1}_V U, Z) = g_1(T_V FU, \mathcal{B}Z) + g_1(\mathcal{H}\nabla^{M_1}_V FU, \mathcal{C}Z).$$

Since ∇^{M_1} is torsion free and $[V, FU] \in \Gamma(ker\pi_*)$ we obtain

$$g_1(\nabla^{M_1}_V U, Z) = g_1(T_V FU, \mathcal{B}Z) + g_1(\nabla^{M_1}_{FU} V, \mathcal{C}Z).$$

Using (3) and (10) we have

$$g_1(\nabla^{M_1}_V U, Z) = g_1(T_V FU, \mathcal{B}Z) + g_1(\nabla^{M_1}_{FU} FV, FCZ)$$

here we have used that μ is invariant. Since π is a conformal submersion, using (12) and Lemma 2.2 we obtain

$$\begin{aligned} g_1(\nabla^{M_1}_V U, Z) \quad &= g_1(T_V FU, \mathcal{B}Z) + \frac{1}{\lambda^2} g_1(\mathcal{H}grad \ln\lambda, FU) g_2(\pi_* FV, \pi_* FCZ) \\ &- \frac{1}{\lambda^2} g_1(\mathcal{H}grad \ln\lambda, FV) g_2(\pi_* FU, \pi_* FCZ) \\ &+ g_1(FU, FV) \frac{1}{\lambda^2} g_2(\pi_*(\mathcal{H}grad \ln\lambda), \pi_* FCZ) \\ &+ \frac{1}{\lambda^2} g_2(\nabla^{\pi}_{FU}\pi_* FV, \pi_* FCZ). \end{aligned}$$

Moreover, using Definition 3.1 and (17), we obtain

$$g_1(\nabla^{M_1}_V U, Z) = g_1(T_V FU, \mathcal{B}Z) + g_1(U,V) g_1(\mathcal{H}grad \ln\lambda, FCZ) + \frac{1}{\lambda^2} g_2(\nabla^{\pi}_{FU}\pi_* FV, \pi_* FCZ)$$

which proves $(i) \Leftrightarrow (ii)$. □

From Theorem 4.3, we deduce the following result.

Theorem 4.4. *Let π be a conformal anti-invariant submersion from a locally product Riemannian manifold (M_1, g_1, F) onto a Riemannian manifold (M_2, g_2). Then any two conditions below imply the three;*

i. *$ker\pi_*$ defines a totally geodesic foliation on M_1*

ii. *λ is a constant on $\Gamma(\mu)$*

iii. $-\frac{1}{\lambda^2} g_2(\nabla^{\pi}_{FU}\pi_* FV, \pi_* FCZ) \quad = g_1(T_V FU, \mathcal{B}Z)$

for $V, U \in \Gamma(ker\pi_*)$ and $Z \in \Gamma\left((ker\pi_*)^{\perp}\right)$.

Proof. For $V, U \in \Gamma(ker\pi_*)$ and $Z \in \Gamma\left((ker\pi_*)^{\perp}\right)$, from Theorem 4.3 we have

$$g_1(\nabla_V^{M_1} U, Z) = g_1(T_V FU, \mathcal{B}Z) + g_1(U, V) g_1(\mathcal{H} grad \ln \lambda, FCZ) + \frac{1}{\lambda^2} g_2(\nabla_{FU}^{\pi} \pi_* FV, \pi_* FCZ).$$

Now, if we have (i) and (iii), then we obtain

$$g_1(U, V) g_1(\mathcal{H} grad \ln \lambda, FCZ) = 0.$$

From above equation, λ is a constant on $\Gamma(\mu)$. Similarly, one can obtain the other assertions. □

If π is a conformal Lagrangian submersion, then (16) implies that $TM_2 = \pi_*(Fker\pi_*)$. Hence we have the following corollary:

Corollary 4.2. *Let π be a conformal Lagrangian submersion from a locally product Riemannian manifold (M_1, g_1, F) onto a Riemannian manifold (M_2, g_2). Then the following assertions are equivalent to each other;*

i. $ker\pi_*$ *defines a totally geodesic foliation on* M_1.

ii. $T_V FU = 0$

for $V, U \in \Gamma(ker\pi_*)$ and $Z \in \Gamma\left((ker\pi_*)^{\perp}\right)$.

Proof. From Theorem 4.3 we have

$$g_1(\nabla_V^{M_1} U, Z) = g_1(T_V FU, \mathcal{B}Z) + g_1(U, V) g_1(\mathcal{H} grad \ln \lambda, FCZ) + \frac{1}{\lambda^2} g_2(\nabla_{FU}^{\pi} \pi_* FV, \pi_* FCZ).$$

for $V, U \in \Gamma(ker\pi_*)$ and $Z \in \Gamma\left((ker\pi_*)^{\perp}\right)$. Since π is a conformal Lagrangian submersion, we get

$$g_1(\nabla_V^{M_1} U, Z) = g_1(T_V FU, \mathcal{B}Z)$$

which shows $(i) \Leftrightarrow (ii)$. □

5. Totally geodesicness of the conformal anti-invariant submersion

In this section, we shall examine the totally geodesicness of a conformal anti-invariant submersion. We give a necessary and sufficient condition for a conformal anti-invariant submersion to be totally geodesic map. Recall that a smooth map π between two Riemannian manifolds is called totally geodesic if $\nabla \pi_* = 0$ [15].

Theorem 5.1. *Let $\pi : (M_1, g_1, F) \to (M_2, g_2)$ is a conformal anti-invariant submersion from an almost product Riemannian manifold (M_1, g_1, F) to a Riemannian manifold (M_2, g_2). π is totally geodesic map if and only if.*

(a) *π is a horizontally homothetic map,*

(b) *$T_U FV = 0$ and $\mathcal{H}\nabla_U^{M_1} FV \in \Gamma(Fker\pi_*)$,*

(c) *$A_Z FV = 0$ and $\mathcal{H}\nabla_Z^{M_1} FV \in \Gamma(Fker\pi)$*

for $Z, W, Z \in \Gamma\left((ker\pi_*)^\perp\right)$ and $U, V \in \Gamma(ker\pi_*)$.

Proof. (a) For any $Z, W \in \Gamma(\mu)$, from Lemma 2.2 we derive

$$(\nabla\pi_*)(Z, W) = Z(\ln\lambda)\pi_* W + W(\ln\lambda)\pi_* Z - g_1(Z, W)\pi_*(grad \ln\lambda).$$

It is obvious that if π is a horizontally homothetic map, it follows that $(\nabla\pi_*)(Z, W) = 0$. Conversely, if $(\nabla\pi_*)(Z, W) = 0$, taking $W = FZ$ in above equation, we get

$$Z(\ln\lambda)\pi_* FZ + FZ(\ln\lambda)\pi_* Z - g_1(Z, FZ)\pi_*(grad \ln\lambda) = 0. \tag{22}$$

Taking inner product in (31) with $\pi_* FZ$, we obtain

$$g_1(grad \ln\lambda, Z)\lambda^2 g_1(FZ, FZ) + g_1(grad \ln\lambda, FZ)\lambda^2 g_1(Z, FZ) - g_1(Z, FZ)\lambda^2 g_1(grad \ln\lambda, FZ) = 0. \tag{23}$$

From (32), λ is a constant on $\Gamma(\mu)$. On the other hand, for $U, V \in \Gamma(ker\pi_*)$, from Lemma 2.2 we have

$$(\nabla\pi_*)(FU, FV) = FU(\ln\lambda)\pi_* FV + FV(\ln\lambda)\pi_* FU - g_1(FU, FV)\pi_*(grad \ln\lambda).$$

Again if π is a horizontally homothetic map, then $(\nabla\pi_*)(FU, FV) = 0$. Conversely, if $(\nabla\pi_*)(FU, FV) = 0$, putting U instead of V in above equation, we derive

$$2FU(\ln\lambda)\pi_* FU - g_1(FU, FU)\pi_*(grad \ln\lambda) = 0. \tag{24}$$

Taking inner product in (33) with $\pi_* FU$ and since π is a conformal submersion, we have

$$g_1(FU, FU)\lambda^2 g_1(grad \ln\lambda, FU) = 0.$$

From above equation, λ is a constant on $\Gamma(Fker\pi_*)$. Thus λ is a constant on $\Gamma\left((ker\pi_*)^\perp\right)$.

(b) For any $U, V \in \Gamma(ker\pi_*)$, using (3) and (12) we have

$$(\nabla\pi_*)(U,V) = \nabla^{\pi}_{U}\pi_*V - \pi_*\left(\nabla^{M_1}_{U}V\right)$$
$$= -\pi_*\left(F\nabla^{M_1}_{U}FV\right).$$

Then from (7) and (8) we arrive at

$$(\nabla\pi_*)(U,V) = -\pi_*\left(FT_UFV + \mathcal{CH}\nabla^{M_1}_{U}FV\right).$$

From above equation, $(\nabla\pi_*)(U,V) = 0$ if and only if

$$\pi_*\left(FT_UFV + \mathcal{CH}\nabla^{M_1}_{U}FV\right) = 0 \tag{25}$$

Since π is non-singular, this implies $T_UFV = 0$ and $\mathcal{H}\nabla^{M_1}_{U}FV \in \Gamma(Fker\pi_*)$.

(c) For $Z \in \Gamma(\mu)$ and $V \in \Gamma(ker\pi_*)$, from (3) and (12) we get

$$(\nabla\pi_*)(Z,V) = \nabla^{\pi}_{Z}\pi_*V - \pi_*\left(\nabla^{M_1}_{Z}V\right)$$
$$= -\pi_*\left(F\nabla^{M_1}_{Z}FV\right).$$

Using (9) and (10) we have

$$(\nabla\pi_*)(Z,V) = \pi_*\left(FA_ZFV + \mathcal{CH}\nabla^{M_1}_{Z}FV\right).$$

Thus $(\nabla\pi_*)(Z,V) = 0$ if and only if

$$\pi_*\left(FA_ZFV + \mathcal{CH}\nabla^{M_1}_{Z}FV\right) = 0.$$

Then, since π is a linear isomorphism between $(ker\pi_*)^{\perp}$ and TM_2, $(\nabla\pi_*)(Z,V) = 0$ if and only if $A_ZFV = 0$ and $\mathcal{H}\nabla^{M_1}_{Z}FV \in \Gamma(Fker\pi_*)$. Thus proof is complete. □

Here we present another result on conformal anti-invariant submersion to be totally geodesic.

Theorem 5.2 *Let π be a conformal anti-invariant submersion from a locally product Riemannian manifold (M_1, g_1, F) onto a Riemannian manifold (M_2, g_2). If π is a totally geodesic map then*

$$\nabla^{\pi}_{Z}\pi_*W_2 = \pi_*\left(F\left(A_ZFW_1 + \mathcal{V}\nabla^{M_1}_{Z}BW_2 + A_ZCW_2\right) + C\left(\mathcal{H}\nabla^{M_1}_{Z}FW_1 + A_ZBW_2 + \mathcal{H}\nabla^{M_1}_{Z}CW_2\right)\right)$$

for any $Z \in \Gamma\left((\ker\pi_*)^{\perp}\right)$ and $W = W_1 + W_2 \in \Gamma(TM)$, where $W_1 \in \Gamma(\ker\pi_*)$ and $W_2 \in \Gamma\left((\ker\pi_*)^{\perp}\right)$.

Proof. Using (3) and (12) we have

$$(\nabla\pi_*)(Z,W)=\nabla^{\pi}_{Z}\pi_*W-\pi_*\left(F\nabla^{M_1}_{Z}FW\right)$$

for any $Z\in\Gamma\left((\ker\pi_*)^{\perp}\right)$ and $W\in\Gamma(TM_1)$. Then from (9), (10) and (15) we get

$$\begin{aligned}(\nabla\pi_*)(Z,W)=\nabla^{\pi}_{Z}\pi_*W_2-\pi_*\big(&FA_ZFW_1+B\mathcal{H}\nabla^{M_1}_{Z}FW_1+C\mathcal{H}\nabla^{M_1}_{Z}FW_1+BA_ZBW_2\\&+CA_ZBW_2+F\mathcal{V}\nabla^{M_1}_{Z}BW_2+FA_ZCW_2+B\mathcal{H}\nabla^{M_1}_{Z}CW_2+C\mathcal{H}\nabla^{M_1}_{Z}CW_2\big)\end{aligned}$$

for any $W=W_1+W_2\in\Gamma(TM)$, where $W_1\in\Gamma(\ker\pi_*)$ and $W_2\in\Gamma\left((\ker\pi_*)^{\perp}\right)$. Thus taking into account the vertical parts, we find

$$\begin{aligned}(\nabla\pi_*)(Z,W)=\nabla^{\pi}_{Z}\pi_*W_2-\pi_*\big(&F\big(A_ZFW_1+\mathcal{V}\nabla^{M_1}_{Z}BW_2+A_ZCW_2\big)\\&+C\big(\mathcal{H}\nabla^{M_1}_{Z}FW_1+A_ZBW_2+\mathcal{H}\nabla^{M_1}_{Z}CW_2\big)\big)\end{aligned}$$

which gives our assertion. □

6. Examples

In this section, we now give some examples for conformal anti-invariant submersions from almost product Riemannian manifolds.

Example 6.1. *Every anti-invariant Riemannian submersion is a conformal anti-invariant submersion with $\lambda=I$, where I is the identity function* [7].

We say that a conformal anti-invariant submersion is proper if $\lambda\neq I$. We now present an example of a proper conformal anti-invariant submersion. Note that given an Euclidean space $\mathbb{R}^4$ with coordinates $(x_1,\dots,x_4)$, we can canonically choose an almost product structure F on $\mathbb{R}^4$ as follows:

$$F\left(a_1\frac{\partial}{\partial x_1}+a_2\frac{\partial}{\partial x_2}+a_3\frac{\partial}{\partial x_3}+a_4\frac{\partial}{\partial x_4}\right)=a_3\frac{\partial}{\partial x_1}+a_4\frac{\partial}{\partial x_2}+a_1\frac{\partial}{\partial x_3}+a_2\frac{\partial}{\partial x_4},$$
$$a_1,\dots,a_4\in\mathbb{R}. \tag{26}$$

Example 6.2. *Let π be a submersion defined by*

$$\pi:\underset{(x_1,x_2,x_3,x_4)}{\mathbb{R}^4}\to\underset{(\cos x_1\sinh x_2,\,\sin x_1\cosh x_2).}{\mathbb{R}^2}$$

Then it follows that

$$ker\pi_* = span\{V_1 = \partial x_3,\ \ V_2 = \partial x_4\}$$

and

$$(ker\pi_*)^{\perp} = span\{X_1 = \partial x_1,\ \ X_2 = \partial x_2\}.$$

Hence, we have $FV_1 = X_1$ *and* $FV_2 = X_2$ *imply that* $F(ker\pi_*) = (ker\pi_*)^{\perp}$. *Also by direct computations, we get*

$$\pi_* X_1 = -\sin x_1 \sinh x_2 \partial y_1 + \cos x_1 \cosh x_2 \partial y_2,$$

$$\pi_* X_2 = \cos x_1 \cosh x_2 \partial y_1 + \sin x_1 \sinh x_2 \partial y_2.$$

Hence, we have

$$g_2(\pi_* X_1, \pi_* X_1) = \left(\sin^2 x_1 \sinh^2 x_2 + \cos^2 x_1 \cosh^2 x_2\right) g_1(X_1, X_1),$$

$$g_2(\pi_* X_2, \pi_* X_2) = \left(\sin^2 x_1 \sinh^2 x_2 + \cos^2 x_1 \cosh^2 x_2\right) g_1(X_2, X_2),$$

where g_1 and g_2 denote the standard metrics (inner products) of $\mathbb{R}^4$ and $\mathbb{R}^2$. Thus π is a conformal anti-invariant submersion with $\lambda^2 = \left(\sin^2 x_1 \sinh^2 x_2 + \cos^2 x_1 \cosh^2 x_2\right)$.

Example 6.3. *Let* π *be a submersion defined by*

$$\pi : \underset{(x_1, x_2, x_3, x_4)}{\mathbb{R}^4} \rightarrow \underset{\left(\frac{e^{x_3}\sin x_4}{\sqrt{2}}, \frac{e^{x_3}\cos x_4}{\sqrt{2}}\right)}{\mathbb{R}^2}.$$

Then it follows that

$$ker\pi_* = span\{V_1 = \partial x_1,\ \ V_2 = \partial x_2\}$$

and

$$(ker\pi_*)^{\perp} = span\{W_1 = \partial x_3,\ \ W_2 = \partial x_4\}.$$

Hence we have $FV_1 = W_1$ *and* $FV_2 = W_2$ *imply that* $F(ker\pi_*) = (ker\pi_*)^{\perp}$. *Also by direct computations, we get*

$$\pi_* W_1 = \frac{e^{x_3}\sin x_4}{\sqrt{2}} \partial y_1 + \frac{e^{x_3}\cos x_4}{\sqrt{2}} \partial y_2,$$

$$\pi_* W_2 = \frac{e^{x_3}\cos x_4}{\sqrt{2}} \partial y_1 - \frac{e^{x_3}\sin x_4}{\sqrt{2}} \partial y_2.$$

Hence, we have

$$g_2(\pi_* W_1, \pi_* W_1) = \left(\frac{e^{x_3}}{\sqrt{2}}\right)^2 g_1(W_1, W_1),$$

$$g_2(\pi_* W_2, \pi_* W_2) = \left(\frac{e^{x_3}}{\sqrt{2}}\right)^2 g_1(W_2, W_2),$$

where g_1 and g_2 denote the standard metrics (inner products) of $\mathbb{R}^4$ and $\mathbb{R}^2$. Thus π is a conformal anti-invariant submersion with $\lambda = \left(\frac{e^{x_3}}{\sqrt{2}}\right)$.

Author details

Mehmet Akif Akyol

Address all correspondence to: mehmetakifakyol@bingol.edu.tr

Department of Mathematics, Faculty of Arts and Sciences, Bingöl University, Bingöl, Turkey

References

[1] O'Neill B. The fundamental equations of a submersion. Michigan Mathematical Journal. 1966;**13**:458-469

[2] Gray A. Pseudo-Riemannian almost product manifolds and submersions. Journal of Mathematics and Mechanics. 1967;**16**:715-737

[3] Watson B. Almost Hermitian submersions. Journal of Differential Geometry. 1976;**11**(1): 147-165

[4] Şahin B. Anti-invariant Riemannian submersions from almost Hermitian manifolds. Central European Journal of Mathematics. 2010;**3**:437-447

[5] Beri A, Erken İK, Murathan C. Anti-invariant Riemannian submersions from Kenmotsu manifolds onto Riemannian manifolds. 2016;**40**:540-552

[6] Cengizhan M, Erken IK. Anti-invariant Riemannian submersions from cosymplectic manifolds onto Riemannian submersions. Univerzitet u Nišu. 2015;**29**(7):1429-1444

[7] Gündüzalp Y. Anti-invariant semi-Riemannian submersions from almost para-Hermitian manifolds. Journal of Function Spaces and Applications. 2013;7. Article ID 720623

[8] Gündüzalp Y. Anti-invariant Riemannian submersions from almost product Riemannian manifolds. Mathematical Science and Applications E-notes. 2013;**1**(1):58-66

[9] Lee JW. Anti-invariant $\xi^{\perp}$-Riemannian submersions from almost contact manifolds. Hacettepe Journal of Mathematics and Statistics. 2013;**42**(3):231-241

[10] Lee JC, Park JH, Şahin B, Song DY. Einstein conditions for the base space of anti-invariant Riemannian submersions and Clairaut submersions. Taiwanese Journal of Mathematics. In press. DOI: 10.11650/tjm.19.2015.5283

[11] Park KS. H-anti-invariant submersions from almost quaternionic Hermitian manifolds. arXiv:1507.04473 [math.DG]

[12] Shahid A, Tanveer F. Anti-invariant Riemannian submersions from nearly Kählerian mani-folds. Univerzitet u Nišu. 2013;**27**(7):1219-1235

[13] Şahin B. Riemannian submersions from almost Hermitian manifolds. Taiwanese Journal of Mathematics. 2012;**17**(2):629-659

[14] Taştan HM. On Lagrangian submersions. Hacettepe Journal of Mathematics and Statistics. 2014;**43**(6):993-1000

[15] Baird P, Wood JC. Harmonic Morphisms between Riemannian Manifolds, London Mathematical Society Monographs, 29. Oxford: Oxford University Press, The Clarendon Press; 2003

[16] Fuglede B. Harmonic morphisms between Riemannian manifolds. Annales De L'Institut Fourier. 1978;**28**:107-144

[17] Ishhara T. A mapping of Riemannian manifolds which preserves harmonic functions. Journal of Mathematics of Kyoto University. 1979;**19**:215-229

[18] Gundmundsson S. The geometry of harmonic morphisms [Ph.D. thesis]. University of Leeds; 1992

[19] Ornea L, Romani G. The fundamental equations of a conformal submersions. Beitrague Z. Algebra and Geometrie/Contributions Algebra and Geometry. 1993; **34**(2):233-243

[20] Akyol MA, Şahin B. Conformal anti-invariant submersions from almost Hermitian manifolds. Turkish Journal of Mathematics. 2016;**40**:43-70

[21] Akyol MA, Şahin B. Conformal semi-invariant submersions. Communications in Contemporary Mathematics. 2017;**19**(2):1650011. DOI: 10.1142/S0219199716500115

[22] Akyol MA. Conformal anti-invariant submersions from cosymplectic manifolds. Hacettepe Journal of Mathematics and Statistics. 2017;**46**(2):177-192

[23] Akyol MA. Conformal semi-invariant submersions from almost product Riemannian manifolds. Acta Mathematica Vietnamica. 2017;**42**(3):491-507. DOI: 10.1007/s40306-016-0193-9

[24] Yano K, Kon M. Structures on Manifolds. Singapore: World Scientific; 1984

[25] Falcitelli M, Ianus S, Pastore AM. Riemannian Submersions and Related Topics. River Edge, NJ: World Scientific; 2004

[26] Urakawa H. Calculus of variations and harmonic maps. American Mathematical Society. 1993;**132**

Chapter 3

Recent Advances of Manifold Regularization

Xueqi Ma and Weifeng Liu

Additional information is available at the end of the chapter

http://dx.doi.org/10.5772/intechopen.79383

Abstract

Semi-supervised learning (SSL) that can make use of a small number of labeled data with a large number of unlabeled data to produce significant improvement in learning performance has been received considerable attention. Manifold regularization is one of the most popular works that exploits the geometry of the probability distribution that generates the data and incorporates them as regularization terms. There are many representative works of manifold regularization including Laplacian regularization (LapR), Hessian regularization (HesR) and p-Laplacian regularization (pLapR). Based on the manifold regularization framework, many extensions and applications have been reported. In the chapter, we review the LapR and HesR, and we introduce an approximation algorithm of graph p-Laplacian. We study several extensions of this framework for pairwise constraint, p-Laplacian learning, hypergraph learning, etc.

Keywords: Laplacian regularization, Hessian regularization, p-Laplacian regularization, semi-supervised learning, manifold learning

1. Introduction

In practical applications, it is generally laborious to obtain the labeled samples, though vast amounts of unlabeled samples are easily achieved and provide auxiliary information. Semi-supervised learning (SSL), which takes the full advantages of unlabeled data, is specifically designed to improve learning performance. In representative semi-supervised learning algorithms, it is usually assumed that the intrinsic geometry of the data distribution is supported on the low-dimensional manifold.

The popular manifold learning methods include principal components analysis (PCA), multidimensional scaling (MDS) [1, 2], generative topological mapping (GTM) [3], locally linear embedding (LLE) [4], ISOMAP [5], Laplacian eigenmaps (LE) [6], Hessian eigenmaps

(HLLE) [7], and local tangent space alignment (LTSA) [8]. PCA aims to find the low-dimensional linear subspace which captures the maximum proportion of the variation within the data. MDS aims to place each object in N-dimensional space such that the between-object distances are preserved as well as possible. GTM can be seen as a nonlinear form of principal component analysis or factor analysis. LLE assumes a given sample can be reconstructed by its neighbors, represents the local geometry and then seeks a low-dimensional embedding. ISOMAP incorporates the geodesic distances imposed by a weighted graph. LE preserves neighbor relations of pairwise samples by manipulations on an undirected weighted graph. HLLE obtains the final low-dimensional representations by applying eigenanalysis to a matrix, which is built by estimating the Hessian over neighborhood. LTSA [8] exploits the local tangent information as a representation of the local geometry, and this local tangent information is then aligned to provide a global coordinate. Regularization is a key idea in the theory of splines [9] and is widely used in machine learning [10] (e.g., support vector machines). In 2006, Belkin et al. [11] proposed the manifold regularization framework by introducing a new regularization term to exploit the geometry of the probability distribution. Based on this framework, many successful manifold regularized semi-supervised learning (MRSSL) algorithms have been reported.

Laplacian regularization (LapR) [11, 12] is one prominent manifold regularization-based SSL algorithm, which approximates the manifold by using the graph Laplacian. Putting the simple calculation and prominent performance together, the LapR-based SSL algorithms have been widely used in many applications. Liu et al. [13] introduced Laplacian regularization for local structure preserving and proposed manifold regularized kernel logistic regression (KLR) for web image annotation. Luo et al. [14] employed manifold regularization to smooth the functions along the data manifold for multitask learning. Ma et al. [15] proposed a local structure preserving method that effectively integrates Laplacian regularization and pairwise constraints for human action recognition. Hu et al. [16] introduced graph Laplacian regularization for joint denoising and superresolution of generalized piecewise smooth images.

Hessian regularization [17] (HesR) has attracted considerable attentions and has shown empirically to perform well in practical problems [18–26]. Liu et al. [27] incorporated both Hessian regularization and sparsity constraints into auto-encoders and proposed a new auto-encoder algorithm called Hessian regularized sparse auto-encoders (HSAE). Liu et al. [28] proposed multi-view Hessian regularized logistic regression for action recognition. While the null space of the graph Laplacian along the underlying manifold is a constant function, HesR steers the learned function varying linearly in reference to the geodesic distance. In result, HesR can be more accurate to describe the underlying manifold of data and achieves the better learning performance than LapR-based ones [18]. However, the stability of Hessian estimation depends mostly on the quality of the local fit for each data point, which leads to inaccurate estimation particularly when the function is heavily oscillating [17].

As a nonlinear generalization of the standard graph Laplacian, discrete p-Laplacian has been well studied in mathematics community and solid properties have been investigated by previous work [29, 30]. Meanwhile, graph p-Laplacian has been proved having the advantages for exploiting the manifold of data distribution. Bühler et al. [31] provided a rigorous proof of the approximation of the second eigenvector of p-Laplacian to the Cheeger cut which indicates the superiority of graph p-Laplacian in local geometry exploiting. Luo et al. [32] proposed full eigenvector analysis of p-Laplacian and obtain a natural global embedding for multi-class clustering problems, instead of using greedy search strategy implemented by previous researchers. Liu et al. [33] proposed p-Laplacian regularized sparse coding for human activity recognition.

In this chapter, we first present some related work, and then introduce several extensions based on the manifold regularization framework. Specifically, we present the approximation of graph p-Laplacian and the p-Laplacian regularization framework.

Notations: We present some notations that will be used throughout this chapter. We use $L^{''}$ as the novel graph Laplacian constructed by the traditional graph Laplacian L and the side information. L_p, L_p^{hp} and $\boldsymbol{L}$ represent the graph p-Laplacian, hypergraph p-Laplacian and ensemble graph p-Laplacian, respectively.

2. Related works

This section reviews some related works on manifold regularization, pairwise constraints and hypergraph learning.

2.1. Manifold regularization framework

In semi-supervised learning, assume that N training samples X containing l labeled samples $\{(x_i, y_i)\}_{i=1}^{l}$ and u unlabeled samples $\{(x_j)\}_{j=l+1}^{l+u}$ are available. The labeled samples are pairs generated from probability distribution, while unlabeled samples are simply drawn according to the marginal distribution. To utilize marginal distribution induced by unlabeled samples, we assume that if two points x_1, x_2 are close in the intrinsic geometry of marginal distribution, then the labels of x_1 and x_2 are similar.

Manifold regularized method introduces appropriate penalty term $(\|f\|_I^2)$ and reproducing kernel Hilbert spaces (RKHS) norm $(\|f\|_K^2)$ that is used to control the complexity of the intrinsic geometric structure of the function and the complexity of the classification model, respectively. By incorporating two regularization terms, the standard framework aims to minimize the following function:

$$f^* = arg \min_{f \epsilon H_K} \frac{1}{l} \sum_{i=1}^{l} V(x_i, y_i, f) + \Upsilon_A \|f\|_K^2 + \Upsilon_I \|f\|_I^2. \tag{1}$$

where V is some loss function, such as the hinge loss function $max[0, 1 - y_i f(x_i)]$ for support vector machines (SVM). The parameters Υ_A and Υ_I balance the loss function and two regularization terms. For semi-supervised learning, the manifold regularization term $\|f\|_I^2$ is a key to smooth function along the manifold estimated from the unlabeled samples.

2.2. Pairwise constraints

Pairwise constraints (side information) [34, 35] is a type of supervised information that specify whether a pair of data samples belong to the same class (must-link constraints) or different classes (cannot-link constraints). Compared with class labels, pairwise constraints can provide us weak and more general supervised information. Currently, it has been widely used in semi-supervised clustering [36, 37], distance metric learning [38], feature selection [39] and dimension reduction [40, 41].

Donate $X = \{x_i\}_{i=1}^{n}$ as data set with $Y = \{y_i\}_{i=1}^{n}$ as class labels. Let $M = \{(x_i, x_j)\}$ be the pairwise must-link constraints set and $C = \{(x_i, x_j)\}$ be the pairwise cannot-link constraints set, that is,

$$\mathrm{M} = \{(x_i, x_j) | x_i \text{ and } x_j \text{ belong to the same class }\}$$

$$\mathrm{C} = \{(x_i, x_j) | x_i \text{ and } x_j \text{ belong to different classes }\}.$$

Defined on the pairwise must-link constraint set and the cannot-link constraint set, we construct similarity matrices S^M and S^C, respectively:

$$S_{ij}^{M} = \begin{cases} 1, & if\,(x_i, x_j) \in M \\ 0, & otherwise \end{cases} \tag{2}$$

$$S_{ij}^{C} = \begin{cases} 1, & if\,(x_i, x_j) \in C \\ 0, & otherwise \end{cases}. \tag{3}$$

Then, the must-link Laplacian matrix L^M is given by $L^M = D^M - S^M$, and the cannot-link Laplacian matrix L^C is given by $L^C = D^C - S^C$. Where D^M and D^C are two diagonal matrices with $D_{ii}^{M} = \sum_{j=1}^{n} S_{ij}^{M}$ and $D_{ii}^{C} = \sum_{j=1}^{n} S_{ij}^{C}$, respectively.

Ding et al. [42] introduced pairwise constraints into spectral clustering algorithm. Especially, they revised the distances between sample points by the distance matrix D, where $D_{ij} = \begin{cases} 0 & if\,(x_i, x_j) \in M \\ \infty & if\,(x_i, x_j) \in C \end{cases}$.

Kalakech et al. [43] developed a semi-supervised constraint score by using both pairwise constraints and local properties of the unlabeled data.

Luo et al. [44] denoted the training set with side information by $\{x_i, x_j, y_{ij}\}_{i,j=1}^{N}$, where $y_{ij} = \pm 1$ indicates x_i and x_j are similar or dissimilar. The side information was utilized by denoting the loss function $y_{ij}\left[1 - \|x_i - x_j\|_{A_m}^2\right]$, where A_m is the metric in the m'th heterogeneous domain.

2.3. Hypergraph learning

Hypergraph [45] is a generalization of a simple graph. Compared with simple graphs, a hypergraph illustrates the complex relationship by hyperedges that connect three or more

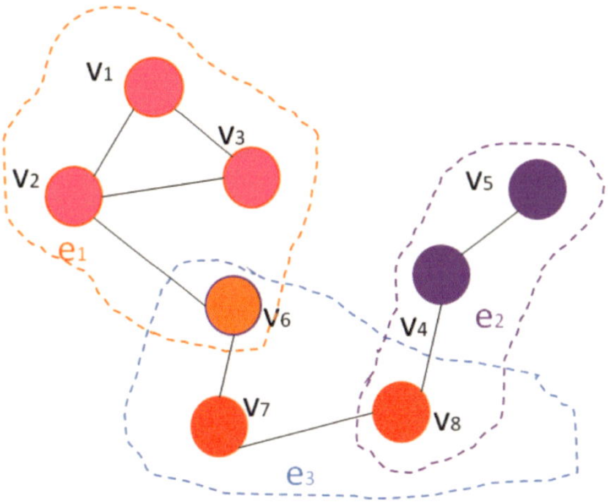

	e1	e2	e3
v1	1	0	0
v2	1	0	0
v3	1	0	0
v4	0	1	0
v5	0	1	0
v6	1	0	1
v7	0	0	1
v8	0	1	1

Figure 1. The block scheme of hypergraph. Left: A simple graph in which two points are joined together by an edge if they are highly similar. A hypergraph completely illustrates the complex relationship among points by hyperedges. Right: The H matrix of the hypergraph. The entry (v_i, e_j) is set to 1 if a hyperedge e_j contains v_i, or 0 otherwise.

vertices (see in **Figure 1**). Thus, the hypergraph contains more local structure information in comparison to simple graph. Hypergraph has been widely used in image classification [46], ranking [47] and video segmentation [48].

Let $G = (V, E)$ denote a hypergraph with the vertex set V and the hyperedge set E. Denote the weight associated with each hyperedge e as $w(e)$. The degree $d(v)$ of a vertex is defined by $d(v) = \sum_{\{e \in E | v \in e\}} w(e)$. The degree of a hyperedge e is denoted as $\delta(e) = |e|$. Denote the vertex-edge incident matrix H by a $|V| \times |E|$ matrix, where entry $h(v, e) = 1$ if $v \in e$, and $h(v, e) = 0$ otherwise. By these definitions, we have:

$$d(v) = \sum_{e \in E} w(e)h(v, e), \delta(e) = \sum_{v \in V} h(v, e). \tag{4}$$

Then, we denote Dv as the diagonal matrices consisting of vertex degree, D_e as the diagonal degree matrices of each hyperedge and W as the diagonal matrix of edge weights. Then, the hypergraph Laplacian can be defined.

A number of different methods have been used in the literature to build the graph Laplacian of hypergraphs. The first category includes star expansion [49], clique expansion [49], Rodriquez's Laplacian [50], etc. These methods aim to construct a simple graph from the original hypergraph, and then partitioning the vertices by spectral clustering techniques. The second category of approaches defines a hypergraph Laplacian using analogies from the simple graph Laplacian. Representative methods in this category include Bolla's Laplacian [51], Zhou' normalized Laplacian [52], etc. According to [52], the normalized hypergraph Laplacian L^{hp} is defined as

$$L^{hp} = I - Dv^{-1/2} HWD_e^{-1} H^T Dv^{-1/2}. \tag{5}$$

It is worth noting that L^{hp} is positive semi-definite. The adjacency matrix of hypergraph W^{hp} can be formulated as follows:

$$W^{hp} = HWH^T - Dv. \tag{6}$$

For a simple graph, the edge degree matrix D_e is replaced by $2I$. Thus, the standard graph Laplacian is

$$\begin{aligned} L &= I - \frac{1}{2} Dv^{-\frac{1}{2}} HWH^T Dv^{-\frac{1}{2}} \\ &= \frac{1}{2}\left(I - Dv^{-1/2} W^{hp} Dv^{-1/2}\right). \end{aligned} \tag{7}$$

3. LapR-based SSL

Laplacian regularization is one of most prominent manifold regularization methods that utilizes the graph Laplacian matrix to characterize the manifold structure. In this section, we introduce the traditional Laplacian support vector machines (LapSVM) and Laplacian kernel least squares (LapKLS) as examples of Laplacian regularization algorithms. Then, we extend the algorithms by building the novel graph Laplacian L'' which combines the traditional graph Laplacian L with the side information to boost locality preservation.

3.1. LapSVM and LapKLS

As previously mentioned, the manifold regularization framework is built by Eq. (1). The traditional LapSVM solves this optimization problem with the hinge loss function

$$f(*) = arg\min_{f \epsilon H_K} \frac{1}{l} \sum_{i=1}^{l} \left(1 - y_i f(x_i)\right)_+ + \Upsilon_A \|f\|_K^2 + \frac{\Upsilon_I}{(l+u)^2} \mathbf{f}^T L \mathbf{f}. \tag{8}$$

where $\mathbf{f}$ is given as $\mathbf{f} = [f(x_1), f(x_2), \cdots, f(x_{l+u})]^T$, L is the graph Laplacian with $L = D - W$, where W_{ij} is weight vector, the diagonal matrix D is given by $D_{ii} = \sum_{j=1}^{n} W_{ij}$.

According to the representer theorem, the solution of the above problem can be expressed as below:

$$f^*(x) = \sum_{i=1}^{l+u} \alpha_i^* \boldsymbol{K}(x_i, x). \tag{9}$$

where $\boldsymbol{K}$ is the kernel function. Therefore, we rewrite the objective function as

$$f(*) = arg\min_{f \epsilon H_K} \frac{1}{l} \sum_{i=1}^{l} \left(1 - y_i f(x_i)\right)_+ + \Upsilon_A \alpha^T \boldsymbol{K} \alpha + \frac{\Upsilon_I}{(l+u)^2} \alpha^T \boldsymbol{K} L \boldsymbol{K} \alpha. \tag{10}$$

By employing the least square loss in Eq. (10), we can present the locality preserved kernel least squares model defined in Eq. (11) as follows

$$f(*) = \min_{f\epsilon H_K} \frac{1}{l}\sum_{i=1}^{l} (y_i - f(x_i))^2 + \Upsilon_A \alpha^T \mathbf{K}\alpha + \frac{\Upsilon_I}{(l+u)^2} \alpha^T \mathbf{K}L\mathbf{K}\alpha. \quad (11)$$

Taking the derivation to the objective functions, we can get the solution of α.

3.2. Pairwise constraints-combined manifold regularization

Assume that samples with the similar features tend to have the similar class labels, combining the Laplacian regularization and pairwise constraints is a good way to exploit the local structure and boost the classification results. Therefore, we introduce the pairwise constraints into traditional LapR. Particularly, we introduce three combination strategies based on experiences. Finally, we present the locality preserved support vector machines and kernel least squares respectively.

According to the definition, we can compute the must-link Laplacian matrix L^M and the cannot-link Laplacian matrix L^C. The first two forms of the combination are defined on the traditional graph Laplacian L and must-link constraints and can be written as

$$L'' = L(L^M + \alpha I) \quad (12)$$

and

$$L'' = L + \alpha L^M \quad (13)$$

respectively, where α is the parameter to balance the weight between the two types of Laplacian matrices.

Based on the cannot-link constraints C, we can compute the similarity matrix S as $S_{ij} = \begin{cases} -1, & if\,(x_i, x_j) \in C \\ 1, & otherwise \end{cases}$. The third form of the combination is defined on the traditional graph Laplacian and pairwise cannot-link constraints and can be written as

$$L'' = L.*S. \quad (14)$$

Actually, there are other combination strategies using both the must-link and cannot-link constraints to get a better result than traditional methods. However, the performance is no better than the result using one only from the experiences. Therefore, we just put these three proposed graph Laplacian into practice.

Introducing the novel graph Laplacian L'' to SVM, we rewrite the learning model as follows:

$$f(*) = arg\min_{f\epsilon H_K} \frac{1}{l}\sum_{i=1}^{l} (1 - y_i f(x_i))_+ + \Upsilon_A \|f\|_K^2 + \frac{\Upsilon_I}{(l+u)^2} \mathbf{f}^T L'' \mathbf{f}. \quad (15)$$

According to the representer theorem, the solution of the above problem can be expressed as below:

$$f^*(x) = \sum_{i=1}^{l+u} \alpha_i^* \mathbf{K}(x_i, x). \tag{16}$$

Therefore, we rewrite the objective function as

$$f(*) = \arg \min_{f \epsilon H_K} \frac{1}{l} \sum_{i=1}^{l} (1 - y_i f(x_i))_+ + \Upsilon_A \alpha^T \mathbf{K} \alpha + \frac{\Upsilon_I}{(l+u)^2} \alpha^T \mathbf{K} L'' \mathbf{K} \alpha. \tag{17}$$

By employing the least square loss in Eq. (17), we can present the locality preserved kernel least squares model defined in Eq. (18) as follows

$$f(*) = \min_{f \epsilon H_K} \frac{1}{l} \sum_{i=1}^{l} (y_i - f(x_i))^2 + \Upsilon_A \Upsilon_A \alpha^T \mathbf{K} \alpha + \frac{\Upsilon_I}{(l+u)^2} \alpha^T \mathbf{K} L'' \mathbf{K} \alpha. \tag{18}$$

We compare our proposed local structure preserving algorithms with the traditional well-known Laplacian algorithms on CAS-YNU-MHAD dataset [53]. CAS-YNU-MHAD dataset contains 10 human actions including jumping up, jumping forward, running, walking S, walking quickly, walking, standing up, sitting down, lying down and typing. **Figure 2** shows the examples. In experiments, we choose the data from four sensors (be placed in the right shoulder, left forearm, left hand and spine) to construct multi-view features. Ninety percent data of per action are randomly selected as the training data, and the rest for testing.

In semi-supervised classification experiments, we randomly select a certain percentage (10, 20, 30, 50%) samples of training data as labeled data. All the classification methods are measured by the average precision (AP) [54] based on the testing data. Note that the supervised information

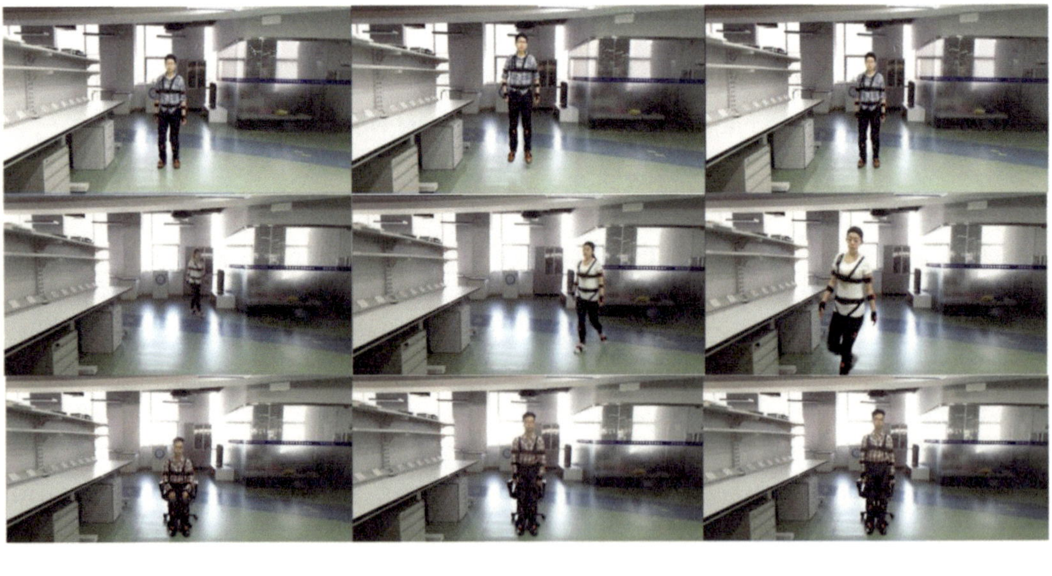

Figure 2. Three examples from 10 actions, jumping up, walking S and sitting down (up to bottom).

http://dx.doi.org/10.5772/intechopen.79383

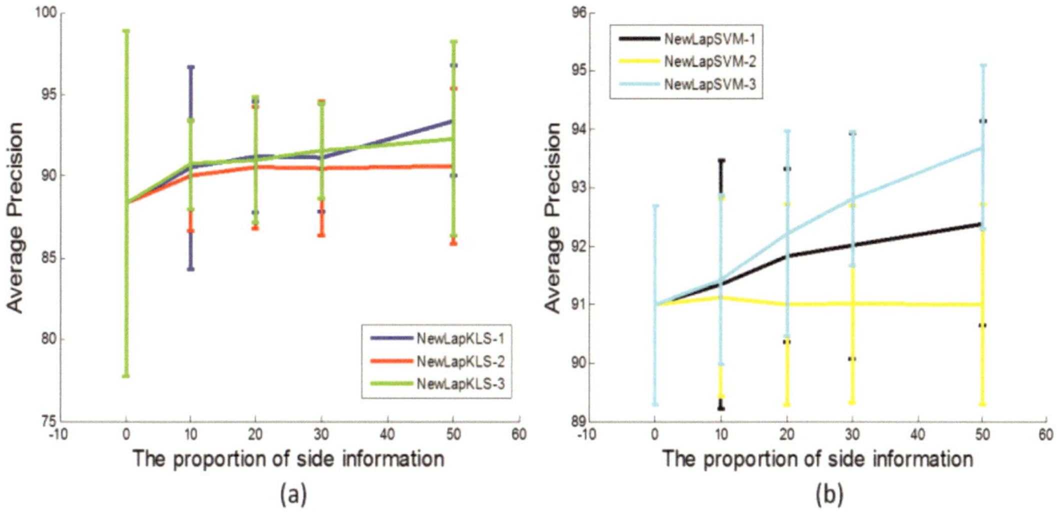

Figure 3. The total classification result under 10% labels (a) KLS, (b) SVM.

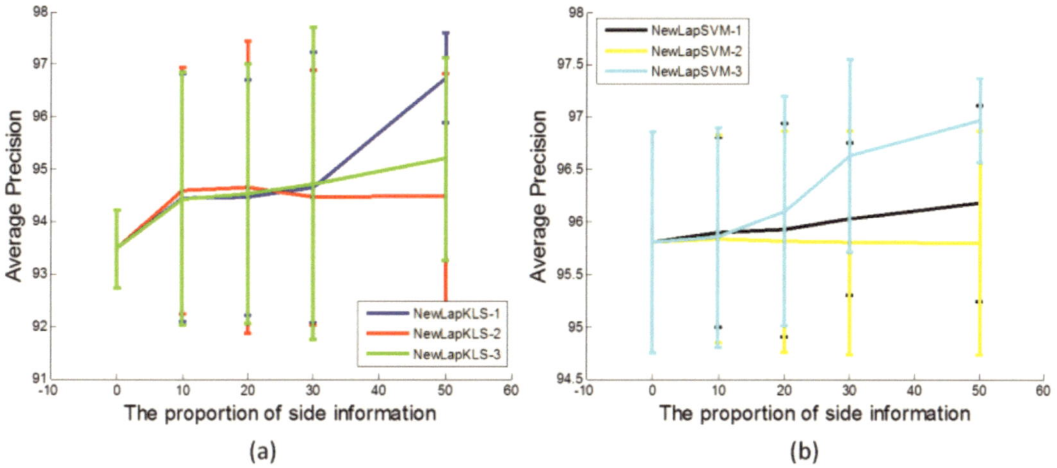

Figure 4. The total classification result under 20% labels (a) KLS, (b) SVM.

(labeled information and side information) are randomly selected from training set. To avoid any potential bias induced by data selecting, the above process is repeated for five times.

For the first two proposed algorithms using the must-link constraints, we first determine the parameter α which balances the traditional graph Laplacian and the must-link Laplacian matrix. The parameter α of novel methods is tuned from the candidate set $\{e^i | i = -10, -9, -8, \cdots, 10\}$ through cross-validation. In addition, the regularization parameters Υ_A, Υ_I are chosen from $\{10^{-8},, 10^{-7},, 10^{-6}, \cdots,, 10^{6},, 10^{7},, 10^{8}\}$ through cross-validation on the training data. We verify the AP performance to select the proper parameters. Note that the parameter α may be different for the same classifier to get the best performance under the different proportion of side information. In results, the legend NewLapKLS-1 represents the kernel least squares classifier using algorithm $L'' = L(L^M + \alpha I)$, NewLapSVM-2 stands for the support vector machines classifier using algorithm $L'' = L + \alpha L^M$, and so on.

Figure 3 shows the classification results achieved by KLS and SVM classifiers under the 10% labeled samples. We can see two main points. First, our proposed three local structure preserving algorithms with pairwise constraints usually get the overall better performances than the well-known semi-supervised methods (LapKLS and LapSVM) without side information. Second, we can clearly see, in most cases, the results gradually become better with the increase of side information. From **Figures 4–6**, we can get the analogous observations for our proposed

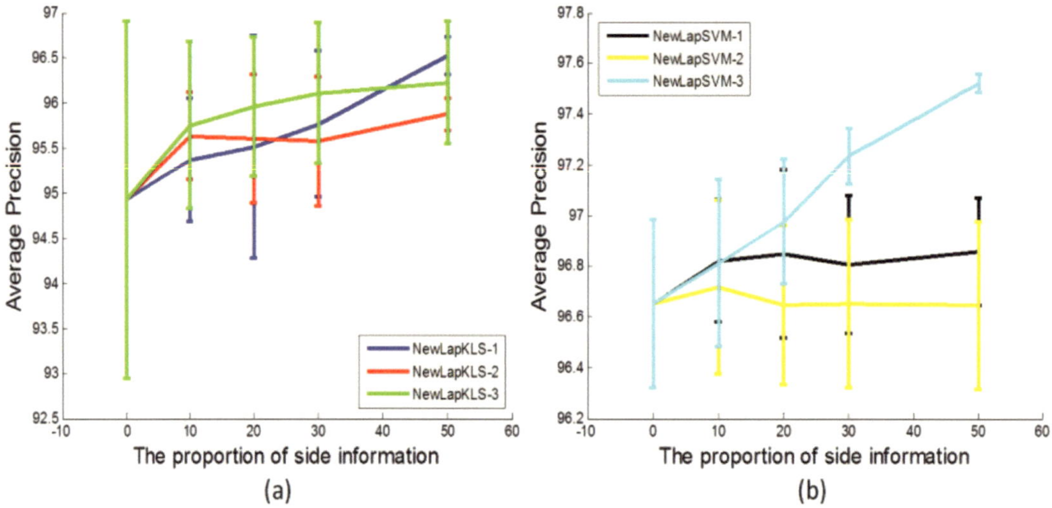

Figure 5. The total classification result under 30% labels (a) KLS, (b) SVM.

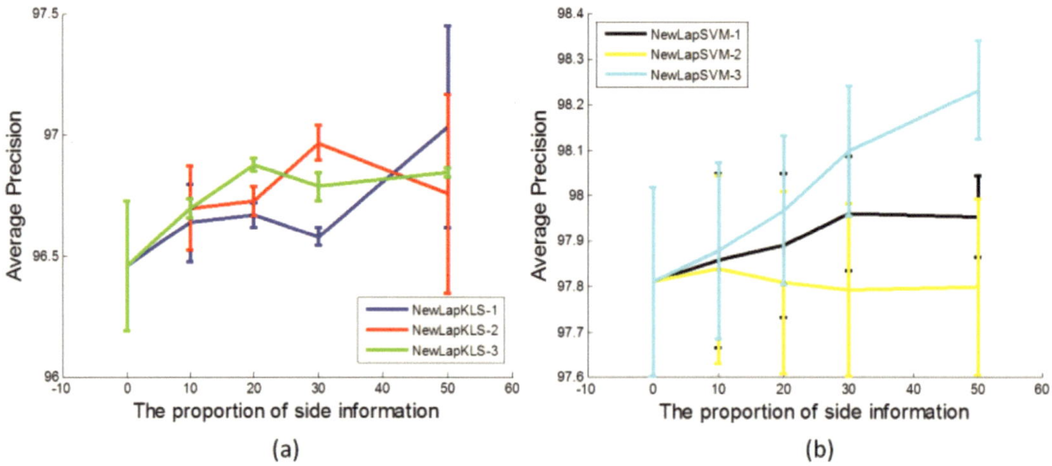

Figure 6. The total classification result under 50% labels (a) KLS, (b) SVM.

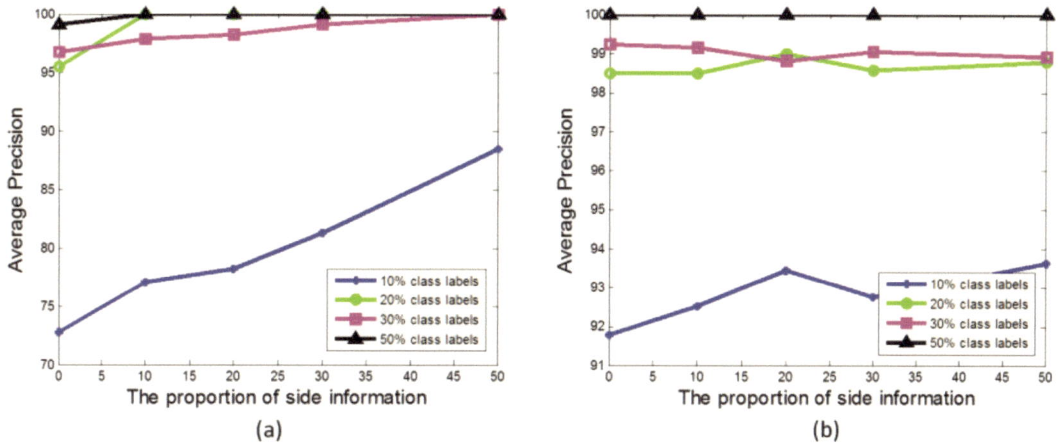

Figure 7. The result of jumping up with the different proportion of side information by LapKLS, LapSVM, NewLapKLS-1 and NewLapSVM-1.

methods compared with their counterparts. These observations indicate that our proposed learning model can better explore and exploit the local structure by taking advantage of the geometrical structure information in the pairwise constraints and manifold regularization. What we can note is that the classification results have slight fluctuation with more side information when the number of class labels is large. These observations suggest it is critical to select parameters for our proposed methods.

To investigate whether the single action of CAS-YNU-MHAD can get the outperformance, we choose jumping up as an example in **Figure 7**. We can find that, our proposed algorithm consistently performs better than the previous algorithm without side information. Especially, we can see, the classification result can get a significant development when the number of labeled samples is limited.

4. HesR-based SSL

Although LapR has received extensive attention, it is observed that the null space of the graph Laplacian along the underlying manifold is a constant function that possibly results in poor generalization. In contrast to Laplacian, Hessian can properly exploit the intrinsic local geometry of the data manifold. In recent works [23–26, 28], HesR based SSL algorithms have been proved to achieve better performance than LapR based ones.

Hessian matrix can be computed by the following four steps.

Step 1: Neighborhood construction. Using k-neighborhood to define neighbors in Euclidean distance for each input point x_i, we get neighborhood matrix N_i.

Step 2: Create local tangent coordinates. Conduct singular value decomposition on neighborhood matrix $N_i = UDV$. The first d columns of V ($V_i = [v_1, v_2, \ldots, v_d]$) mean the tangent coordinates of data points x_i.

Step 3: Build local Hessian estimator. Apply Gram-Schmidt procedure on the matrix $[1, V_i, Q_i]$ with the first column is a vector of ones, $Q_i = [v_i \boxtimes v_j]_{1 \le i \le j \le d}$ is a matrix of $m(m+1)/2$ columns to get $\widehat{M}_i^k$. Then taking the last $m(m+1)/2$ columns of $\widehat{M}_i^k$ as H^i.

Step 4: Construct Hessian matrix *H*. A symmetric matrix H is constructed with the entry $H_{ij} = \sum_l \sum_r H^l_{r,i} H^l_{r,j}$.

The HesR model can be expressed in:

$$f(*) = \arg\min_{f \epsilon H_K} \frac{1}{l} \sum_{i=1}^{l} V(x_i, y_i, f) + \Upsilon_A \|f\|_K^2 + \frac{\Upsilon_I}{(l+u)^2} \mathbf{f}^T H \mathbf{f}. \tag{19}$$

Hessian has been widely utilized in improving the SSL classification performance. Liu et al. [18] present multi-view Hessian discriminative sparse coding (mHDSC) which seamlessly integrates Hessian regularization with discriminative sparse coding for multi-view learning problems. In [24], HesR was employed into support vector machine to boost the classifier. In [19], HesR was integrated into multi-view learning for image annotation, extensive experiments on the PASCAL VOC'07 dataset validate the effectiveness of HesR by comparing it with LapR.

5. pLapR-based SSL

Although the *p*-Laplacian has nice theoretical foundations, it is still a strenuous work to approximate graph *p*-Laplacian, which extremely limits the applications of *p*-Laplacian regularization. In this section, we provide an effect and efficient fully approximation of graph *p*-Laplacian, which significantly lows down the computation cost. Then we integrate the approximated graph *p*-Laplacian into manifold regularization framework and develop *p*-Laplacian regularization. Based on the pLapR, several extended algorithms were proposed.

5.1. pLapR

The graph *p*-Laplacian is approximated by getting all eigenvectors and eigenvalues of *p*-Laplacian [55]. Assume that $f^{*1}, f^{*2}, \cdots, f^{*K}$ are *K* eigenvectors of *p*-Laplacian Δ_p^w associated with unique eigenvalues $\lambda_1^*, \lambda_2^*, \cdots, \lambda_K^*$. Luo et al. [32] introduced an approximation for full eigenvectors of *p*-Laplacian by solving the following *p*-Laplacian embedding problem:

$$\min \mathcal{F}J_E(\mathcal{F}) = \sum_k \frac{\sum_{ij} w_{ij} \left| f_i^k - f_j^k \right|^p}{\|f^k\|_p^p} \qquad s.t. \quad \mathcal{F}^T \mathcal{F} = I. \tag{20}$$

Solving the Eq. (20) with the gradient descend optimization, we can then obtain the full eigenvalues $\Lambda^* = (\lambda_1^*, \lambda_2^*, \cdots, \lambda_K^*)$ of *p*-Laplacian associated with the eigenvectors $\mathcal{F}^* = (f^{*1}, f^{*2}, \cdots, f^{*K})$ by $\lambda_p = \frac{\sum_{ij} w_{ij} |f_i - f_j|^p}{2\|f\|_p^p}$. Finally, the graph *p*-Laplacian approximated by $L_p = \mathcal{F}^* \Lambda \mathcal{F}^{*T}$.

We introduce the approximation graph p-Laplacian into a regularizer to exploit the intrinsic local geometry of the data manifold. Therefore, in p-Laplacian regularization framework, the optimization problem in Eq. (1) becomes

$$f^* = \arg\min_{f \epsilon H_K} \frac{1}{l} \sum_{i=1}^{l} V(x_i, y_i, f) + \Upsilon_A \|f\|_K^2 + \frac{\Upsilon_I}{(l+u)^2} \mathbf{f}^T L_p \mathbf{f}. \tag{21}$$

Here, L_p is the graph p-Laplacian.

The proposed pLapR can be applied to variant MRSSL-based applications with different choices of loss function. Here, we apply pLapR to support vector machines (SVM) and kernel least squares (KLS) as examples.

Applying the hinge loss function in p-Laplacian learning, the p-Laplacian support vector machines (pLapSVM) solves the following optimization problem:

$$f^* = \arg\min_{f \epsilon H_K} \frac{1}{l} \sum_{i=1}^{l} (1 - y_i f(x_i))_+ + \Upsilon_A \|f\|_K^2 + \frac{\Upsilon_I}{(l+u)^2} \mathbf{f}^T L_p \mathbf{f}. \tag{22}$$

The representer theorem has been proved exist and has the general form in Eq. (16). Hence the optimization problem (21) can be expressed as

$$f^* = \arg\min_{f \epsilon H_K} \frac{1}{l} \sum_{i=1}^{l} (1 - y_i f(x_i))_+ + \Upsilon_A \alpha^T \boldsymbol{K} \alpha + \frac{\Upsilon_I}{(l+u)^2} \alpha^T \boldsymbol{K} L_p \boldsymbol{K} \alpha. \tag{23}$$

We outline the KLS with p-Laplacian regularization. For p-Laplacian kernel least squares (pLapKLS), it solves the following optimization problem

$$f^* = \min_{f \epsilon H_K} \frac{1}{l} \sum_{i=1}^{l} (y_i - f(x_i))^2 + \Upsilon_A \alpha^T \boldsymbol{K} \alpha + \frac{\Upsilon_I}{(l+u)^2} \alpha^T \boldsymbol{K} L_p \boldsymbol{K} \alpha. \tag{24}$$

To evaluate the effectiveness of the proposed pLapR, we apply pLapSVM and pLapKLS to scene recognition on the Scene 67 database [56] and Scene 15 data set [57]. **Figure 8** illustrates the framework of pLapR for scene recognition.

The Scene 67 data set contains 15,620 indoor scene images collected from different sources including online image search tools, online photo sharing sites and the LabelMe dataset. Particularly, these images can be categorized into 67 classes covering 5 big scene groups (i.e., stores, home, public spaces, leisure and working place). Some example images are shown in **Figure 9**.

Scene 15 data set is composed of 15 scene categories, totally 4485 images. Each category has 200–400 images. The images contain not only indoor scenes, such as living room, kitchen, and store, but also outdoor scenes, such as forest, mountain, tall building, open country, and so on (see in **Figure 10**).

For Scene 67 dataset, we randomly select 80 images of each class to form the training set and the rest as testing set. For Scene 15 dataset, 100 images per class are randomly selected as the

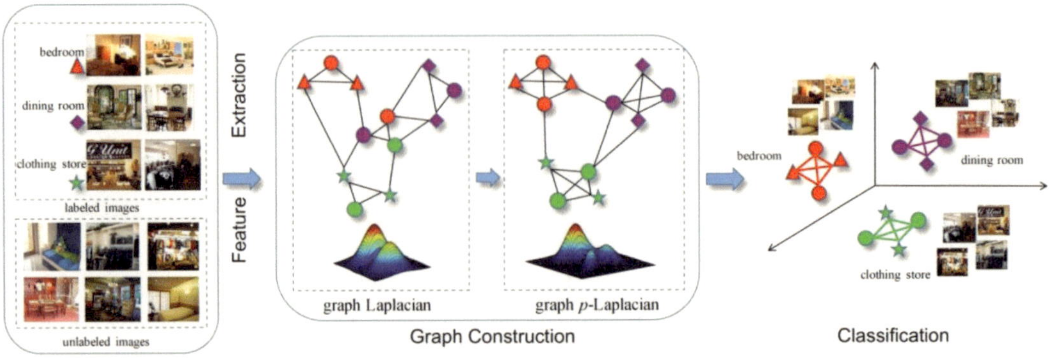

Figure 8. The framework of pLapR for indoor scene recognition.

Figure 9. Some example images of Scene 67 database. The dataset totally has 67 indoor scene categories that can be grouped into 5 big scene groups. Each row demonstrates one big scene group.

training set, and the rest for testing. In semi-supervised experiments, a certain percentage (10, 20, 30, 50%) samples of training set are randomly assigned as labeled data. To avoid any bias introduced by the random partitioning of samples, the above assignment is carried out for five times independently.

Figure 10. Some example images of Scene 15 data set. The dataset totally has 15 scene categories.

The regularization parameters that is, γ_A and γ_I are tuned from the candidate set $\{10^i|i = -10, -9, -8, \cdots, 10\}$ and the parameter p for pLapR from the candidate set $\{1, 1.1, 1.2, \cdots, 3\}$ through cross-validation on the training data with 10% labeled sample, respectively. The performance is measured by the average precision (AP) for single class and mean average precision (mAP) for overall classes. Firstly, we show the mAP boxplot of the pLapR on Scene 67 dataset when $p = 2$ and the standard LapR for comparison in **Figure 11**. We can clearly see that the performance of pLapR with $p = 2$ is similar to standard LapR, which demonstrates that the graph p-Laplacian with $p = 2$ becomes the standard graph Laplacian.

Figure 12 illustrates the performance of pLapKLS with different p values. The upper subfigure is the performance of the Scene 67 database. We observe that the best performance of indoor scene classification on the Scene 67 dataset can be obtained with $p = 1.1$. The lower subfigure is the performance of the Scene 15 database and the best performance is achieved when $p = 1$.

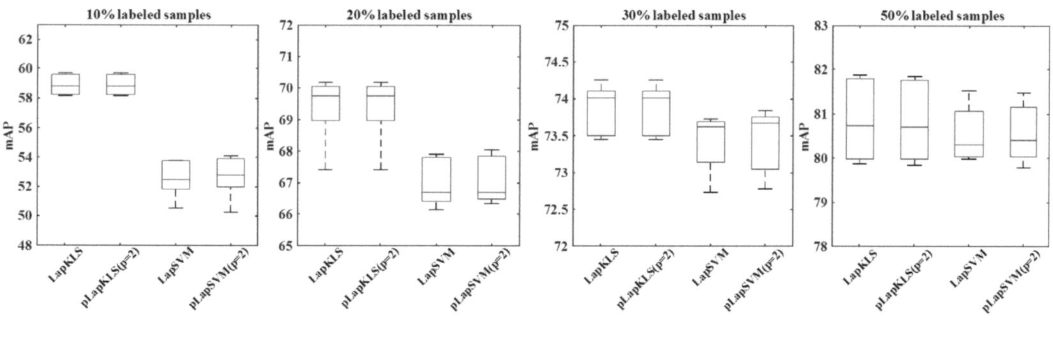

Figure 11. mAP of pLapR($p = 2$) and LapR on Scene 67 dataset. Each subfigure reports the results under different labeled samples. In each subfigure, the y-axis is the mAP over all scene classes, and the x-axis is different classifiers.

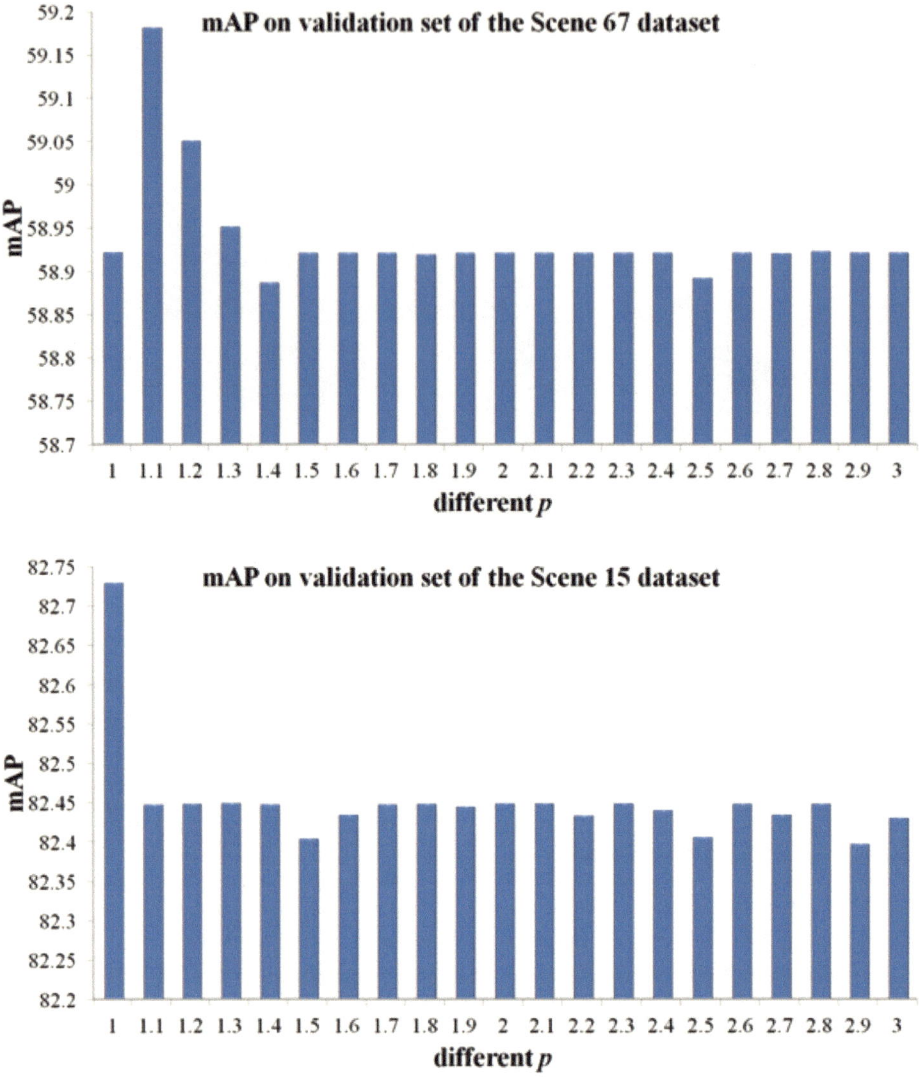

Figure 12. mAP results of pLapKLS under different p with 10% labeled sample.The y-axis is the mAP over all classes, and the x-axis is the parameter p.

Then we evaluate the performance of the pLapR with the representative LapR and HesR. **Figure 13** and **Figure 14** show the mAP performance on Scene 67 data set and Scene 15 data set, respectively. The four subfigures of upper row are KLS methods, and the lower four ones are SVM methods. From the results of two data sets, we can see that the pLapR outperforms both LapR and HesR especially when only a small number of samples labeled.

To discuss the AP performance of different algorithms for single class, we show the results of several classes of Scene 15 data set including mountain, open country, tall building and industrial. Each subfigure corresponds on single scene class. The upper four subfigures are

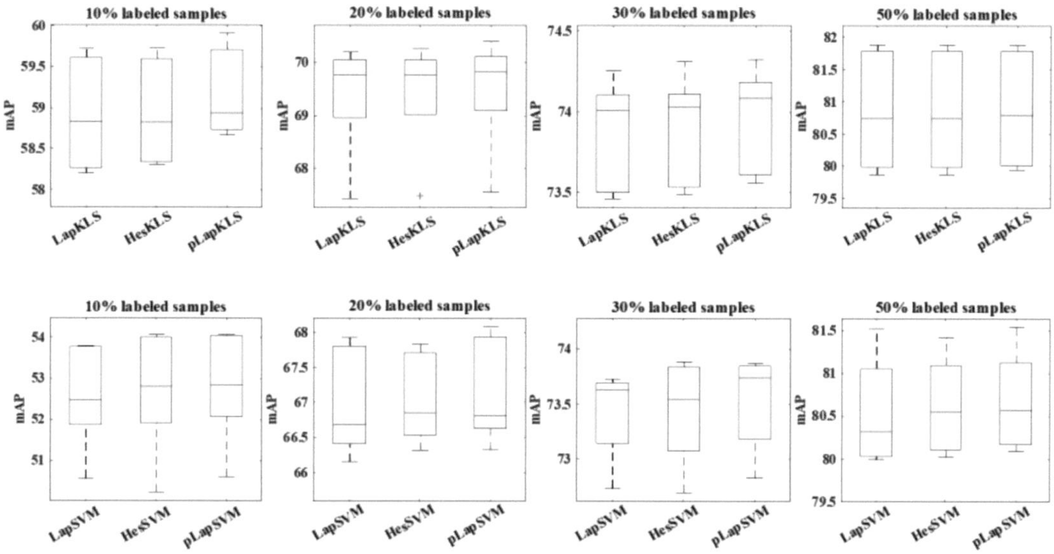

Figure 13. mAP of different algorithms on Scene 67 data set. The four subfigures of upper row are KLS methods, and the lower four ones are SVM methods.

KLS methods, and the lower four ones are SVM methods. In each subfigure, the y-axis is the AP results and the x-axis is the number of labeled samples. From the AP results, we can find that, in most cases, the pLapR performs better than the traditional methods including LapR and HesR (**Figure 15**).

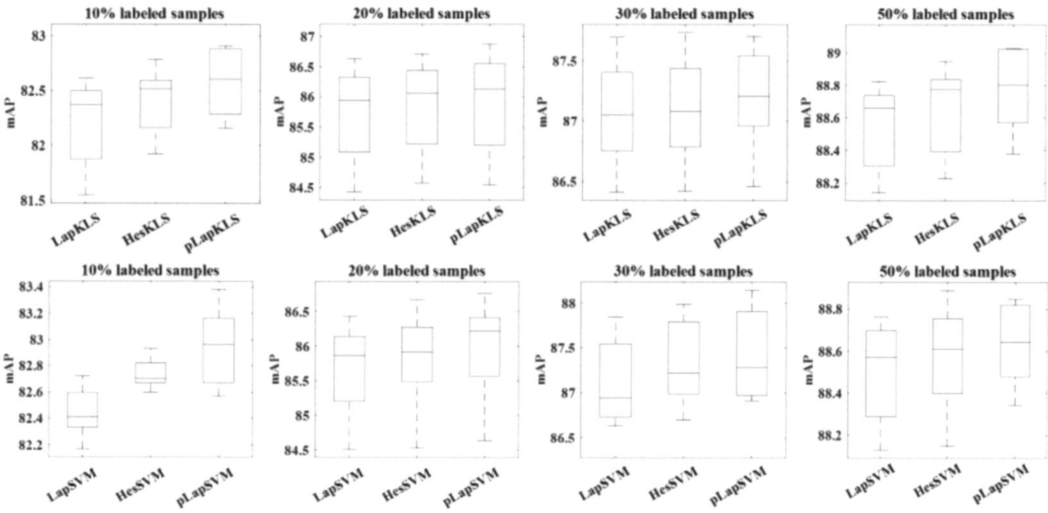

Figure 14. mAP of different algorithms on Scene 15 data set. The four subfigures of upper row are KLS methods, and the lower four ones are SVM methods.

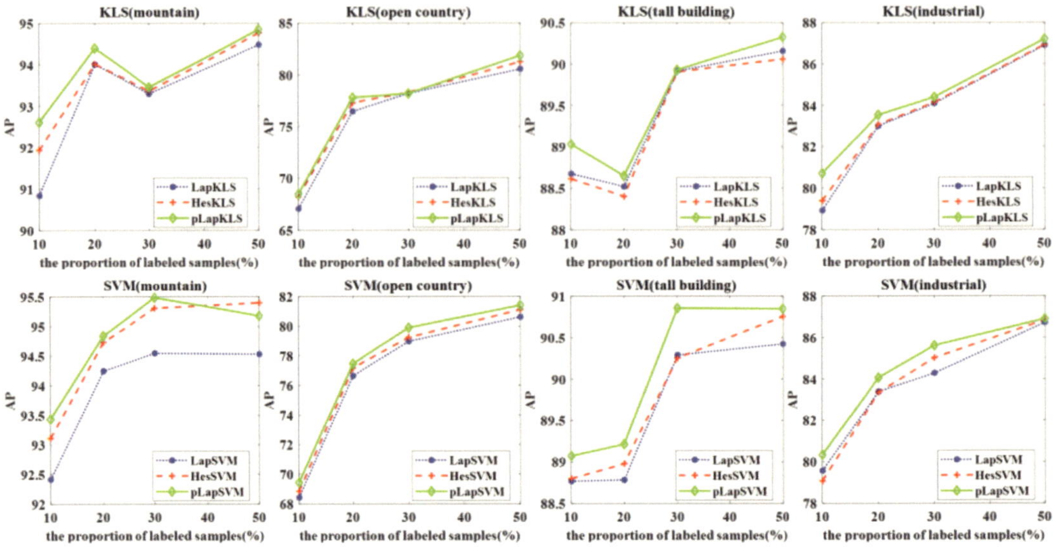

Figure 15. AP of different methods on several classes of Scene 15 data set including mountain, open country, tall building and industrial. Each subfigure corresponds on single scene class. The upper four subfigures are KLS methods, and the lower four ones are SVM methods. In each subfigure, the y-axis is the AP results and the x-axis is the number of labeled samples.

5.2. Hypergraph p-Laplacian (HpLapR)

In this subsection, we propose a hypergraph p-Laplacian regularized method for image recognition. The hypergraph and p-Laplacian [31, 58, 59] both provide convincing theoretical evidence to better preserve the local structure of data. However, the computation of hypergraph p-Laplacian is difficult. We provide an effect and efficient approximation algorithm of hypergraph p-Laplacian. Considering the higher order relationship of samples, the hypergraph p-Laplacian regularizer is built for preserving local structures. Hypergraph p-Laplacian regularization (HpLapR) is also introduced to logistic regression for remote sensing image recognition.

Assume that hypergraph p-Laplacian has n eigenvectors $\mathcal{F}^{*hp} = \left(f^{*hp1}, ,f^{*hp2}, \cdots, ,f^{*hpn}\right)$ associated with unique eigenvalues $\lambda^{*hp} = \left(\lambda_1^{*hp}, \lambda_2^{*hp}, \cdots, \lambda_n^{*hp}\right)$, we compute the approximation of hypergraph p-Laplacian L_p^{hp} by $L_p^{hp} = \mathcal{F}^{*hp}\lambda^{*hp}\mathcal{F}^{*hp^T}$. Thus, it is important to obtain all eigenvectors and eigenvalues of hypergraph p-Laplacian.

Although a complete analysis of hypergraph p-Laplacian is challenging, we can easily generate a hypergraph with a group of hyperedges [52]. In detail, we construct hypergraph Laplacian L^{hp} and compute adjacency matrix W^{hp} by Eq. (5) and Eq. (6), respectively.

Following the study on plapR [31, 55], eigenvalue and the corresponding eigenvector on hypergraph p-Laplacian can be computed by the following hypergraph p-Laplacian embedding problem:

$$\min \mathcal{F}^{hp} J_E(\mathcal{F}^{hp}) = \sum_k \frac{\sum_{ij} w_{ij}^{hp} \left| f_i^{hpk} - f_j^{hpk} \right|^p}{\|f^{hpk}\|_p^p} \quad s.t. \quad \mathcal{F}^{hp^T} \mathcal{F}^{hp} = I \tag{25}$$

Solving the problem of Eq. (25) with the gradient descend optimization. We can also get the full eigenvalue $\lambda^{hp} = \left(\lambda_1^{hp}, \lambda_2^{hp}, \cdots, \lambda_n^{hp}\right)$ by $\lambda_k^{hp} = \frac{\sum_{ij} w_{ij}^{hp} \left| f_i^{hpk} - f_j^{hpk} \right|^p}{\|f^{hpk}\|_p^p}$.

Finally, the approximation of L_p^{hp} can be solved by $L_p^{hp} = \mathcal{F}^{hp} \lambda^{hp} \mathcal{F}^{hp^T}$.

According to the manifold regularization framework, the proposed HpLapR can be written as the following optimization problem:

$$f^* = \arg\min_{f \epsilon H_K} \frac{1}{l} \sum_{i=1}^{l} V(x_i, y_i, f) + \Upsilon_A \|f\|_K^2 + \frac{\Upsilon_I}{(l+u)^2} \mathbf{f}^T L_p^{hp} \mathbf{f}. \tag{26}$$

Here, L_p^{hp} is hypergraph p-Laplacian. We employ the proposed HpLapR with logistic regression. Substitute logistic loss function in Eq. (26), the HpLapR can be rewritten as

$$f^* = \arg\min_{f \epsilon H_K} \frac{1}{l} \sum_{i=1}^{l} \left(\log\left(1 + e^{-y_i f(x_i)}\right)\right) + \Upsilon_A \|f\|_K^2 + \frac{\Upsilon_I}{(l+u)^2} \mathbf{f}^T L_p^{hp} \mathbf{f}^T. \tag{27}$$

According to the representer theorem, the solution of (27) w.r.t. f exists and can be expressed by Eq. (16). Thus, we finally construct the HpLapR as the following optimization problem:

$$f^* = \arg\min_{f \epsilon H_K} \frac{1}{l} \sum_{i=1}^{l} \left(\log\left(1 + e^{-y_i \boldsymbol{K}(x_i, x)\alpha}\right)\right) + \Upsilon_A \alpha^T \boldsymbol{K} \alpha + \frac{\Upsilon_I}{(l+u)^2} \alpha^T \boldsymbol{K} L_p^{hp} \boldsymbol{K} \alpha. \tag{28}$$

Apply the conjugate gradient algorithm, we can get the solution of the optimized f.

To evaluate the effectiveness of the proposed HpLapR, we compare HpLapR with other local structure preserving algorithms including LapR, HLapR and pLapR. **Figure 16** illustrates the framework of HpLapR for UC-Merced data set.

UC-Merced data set [60] consists of totally 2100 land-use images collected from aerial orthoimage with the pixel resolution of one foot. These images were manually selected into 21 classes: agricultural, airplane, baseball diamond, beach, buildings, chaparral, dense residential, forest, freeway, golf course, harbor, intersection, medium density residential, mobile home park, overpass, parking lot, river, runway, sparse residential, storage tanks, and tennis courts (see in **Figure 17**).

In our experiments, we extract high-level visual features using the deep convolution neural network (CNN) [61]. We randomly choose 50 images per class as training samples and the rest

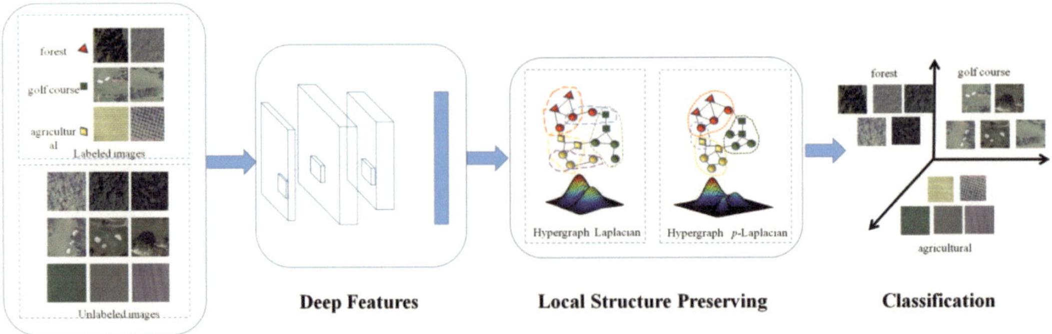

Figure 16. The framework of HpLapR for remote sensing image classification.

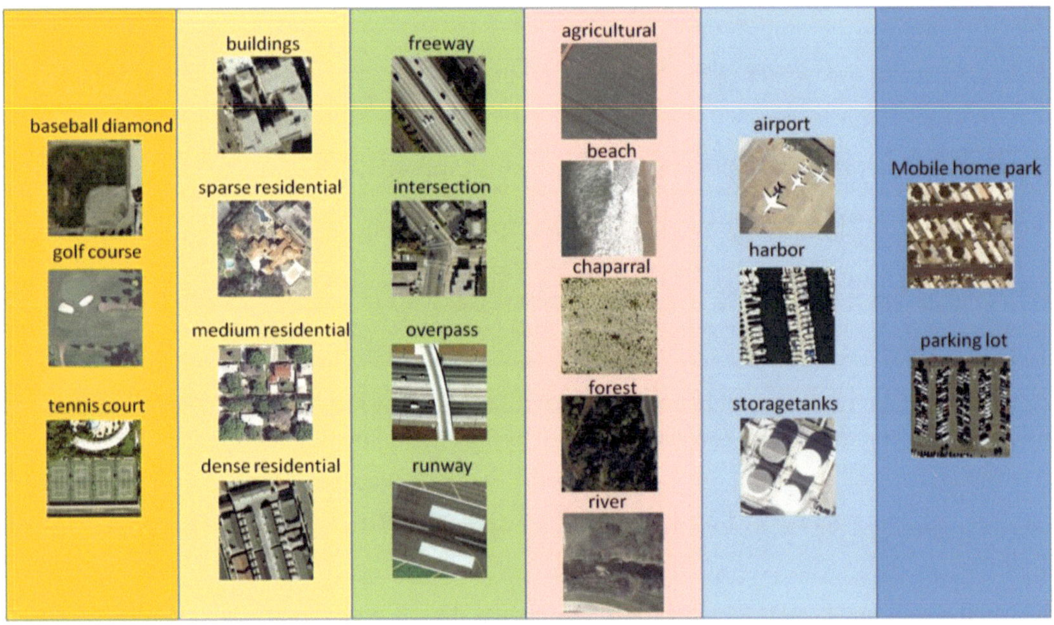

Figure 17. Class examples of UC-Merced data set. The dataset totally has 21 remote sensing categories that can be simply grouped into six groups according to the distinction of land use. Each column represents one group.

as testing samples. For hypergraph construction, we regard each sample in the training set as a vertex, and generate a hyperedge for each vertex with its k nearest neighbors (so the hyperedge connects $k + 1$ samples) [62]. It is worthy to notice that, for our experiments, the kNN-based hyperedges generating method is implemented only in six groups, not in the overall training samples. For example, for a sample of baseball diamond, the vertices of the corresponding hyperedge are chosen from the first group (baseball diamond, golf course and tennis courts) of **Figure 17**. The setting of class labels is as same as pLapR.

We conduct the experiments on the data set to obtain the proper modal parameters. The neighborhood size k of a hypergraph varies in a range $\{5, 6, 7, \cdots, 15\}$ through cross-validation. The setting of regularization parameters γ_A, γ_I and p are as same as pLapR experiments.

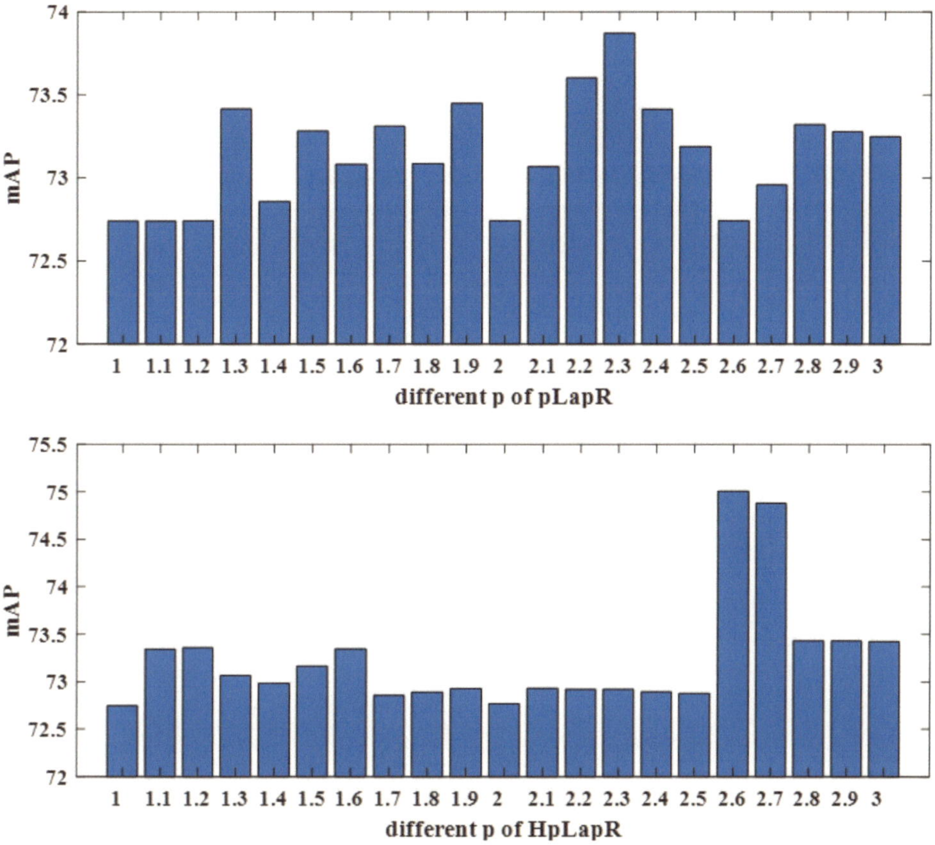

Figure 18. Performance of mAP with different p values on validation set.

Figure 18 illustrates the mAP performance of pLapR and HpLapR on the validation set when p varies. The x-axis is the parameter p and the y-axis is mAP for performance measure. We can see that the best mAP performance for pLapR can be obtained when $p = 2.3$, while the best performance of HpLapR is achieved when $p = 2.6$.

We compare our proposed HpLapR with the representative LapR, HLapR and pLapR. From **Figure 19**, we can observe that, HpLapR outperforms other methods especially when only a small number of samples are labeled. This suggests that our proposed method has the superiority to preserve the local structure of the data because it integrates hypergraph learning with graph p-Laplacian. To evaluate the effectiveness of HpLapR for single class, **Figure 20** shows the AP results of different methods on several land-use classes including beach, dense residential, freeway and tennis court. From **Figure 20**, we can find that, in most cases, HpLapR performs better than both pLapR and HLapR, while pLapR and HLapR consistently outperforms than LapR.

5.3. Ensemble p-Laplacian regularization (EpLapR)

As a natural nonlinear generalization of graph Laplacian, p-Laplacian has been proved having the rich theoretical foundations to better preserve the local structure. However, it is difficult to

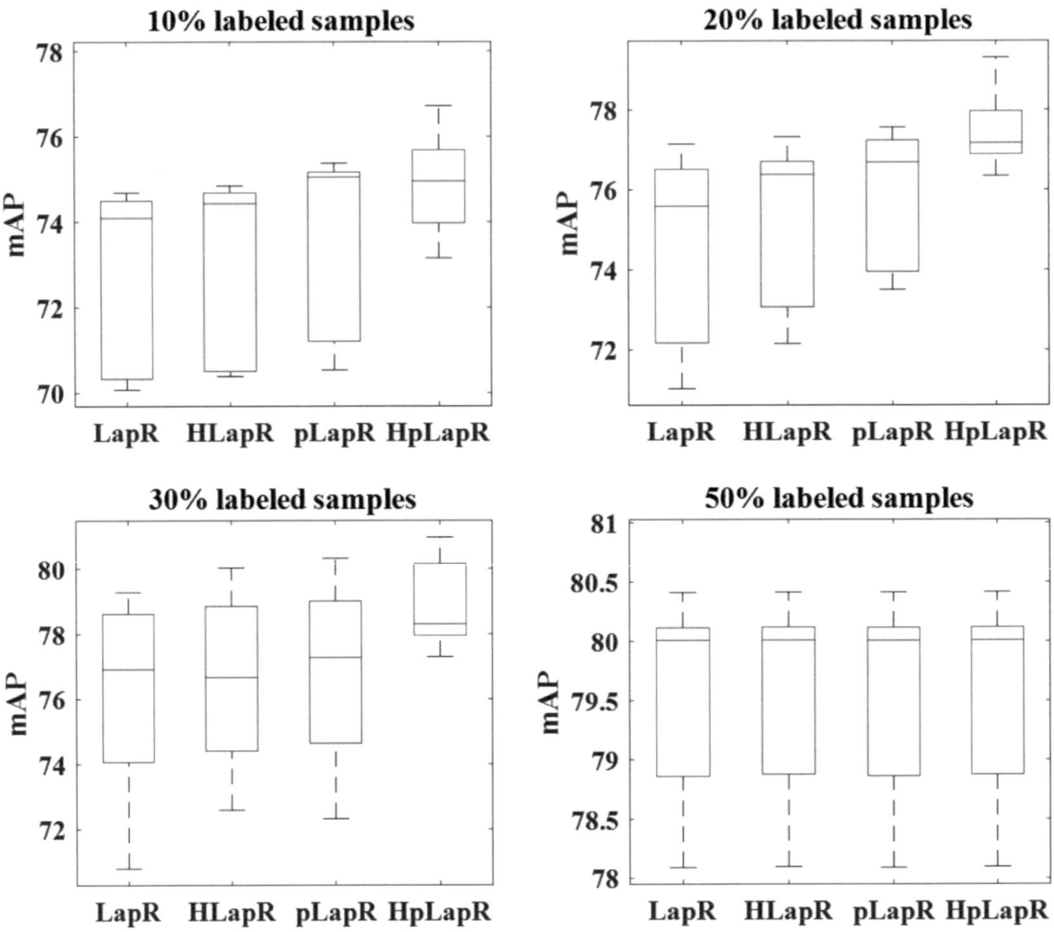

Figure 19. mAP performance of different algorithms.

determine the fitting graph p-Lapalcian, that is, the parameter p that is a critical factor for the performance of graph p-Laplacian. In this section, we develop an ensemble p-Laplacian regularization to fully approximate the intrinsic manifold of the data distribution. EpLapR incorporates multiple graphs into a regularization term in order to sufficiently explore the complementation of graph p-Laplacian. Specifically, we construct a fused graph by introducing an optimization approach to assign suitable weights on different p-value graphs. Then, we conduct semi-supervised learning framework on the fused graph.

Assume a set of candidate graph p-Laplacian $\{L_1^p, \cdots, L_m^p\}$, according to the manifold regularization framework, the proposed EpLapR can be written as the following optimization problem:

$$f^* = \arg\min_{f \epsilon H_K} \frac{1}{l} \sum_{i=1}^{l} V(x_i, y_i, f) + \gamma_A \|f\|_K^2 + \frac{\gamma_I}{n^2} \mathbf{f}^T \mathbf{L}\mathbf{f}. \tag{29}$$

where $\boldsymbol{L}$ is the optimal fused graph with $\boldsymbol{L} = \sum_{k=1}^{m} \mu_k L_k^p,\ s.t. \sum_{k=1}^{m} \mu_k = 1,\ \ \mu_k \geq 0,\ \ for\ k = 1, \cdots, m.$

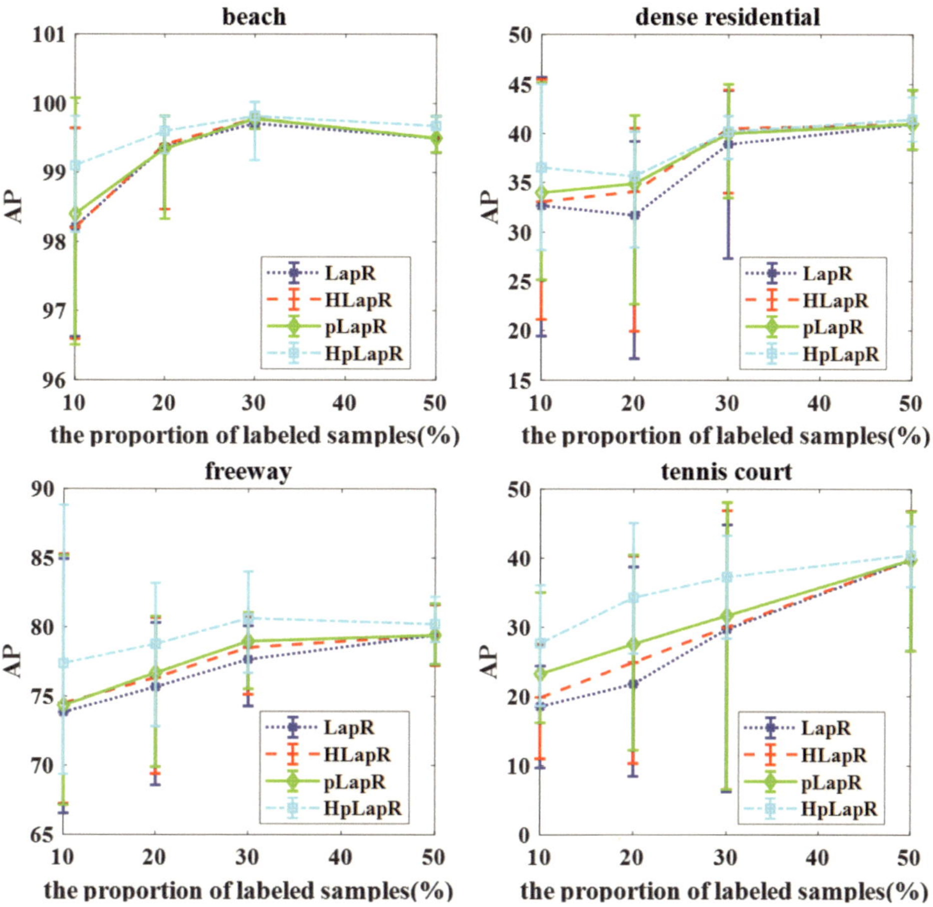

Figure 20. AP performance of different methods on several classes.

To avoid the parameter μ_k overfitting to one graph [63], we make a relaxation by changing μ_k to μ_k^γ, and obtain the optimization problem as:.

$$f^* = arg\,min_{f\epsilon H_K} \frac{1}{l}\sum_{i=1}^{l} V(x_i, y_i, f) + \Upsilon_A \|f\|_K^2 + \frac{\Upsilon_I}{n^2}\mathbf{f}^T\left(\sum_{k=1}^{m}\mu_k^\gamma L_k^p\right)\mathbf{f}. \quad \text{s.t.}\sum_{k=1}^{m}\mu_k = 1,\ \ \mu_k \geq 0,\ \ \text{for } k = 1, \cdots, m \tag{30}$$

The representor theorem presents us with the existence and the general form of Eq. (16) under a fixed μ. Therefore, we rewrite the objective function as

$$f^* = arg\,min_{f\epsilon H_K} \frac{1}{l}\sum_{i=1}^{l} V(x_i, y_i, f) + \Upsilon_A \alpha^T \mathbf{K}\alpha + \frac{\Upsilon_I}{(l+u)^2}\alpha^T\mathbf{K}\left(\sum_{k=1}^{m}\mu_k^\gamma L_k^p\right)\mathbf{K}\alpha. \quad \text{s.t.}\sum_{k=1}^{m}\mu_k = 1,\ \ \mu_k \geq 0,\ \ \text{for } k = 1, \cdots, m \tag{31}$$

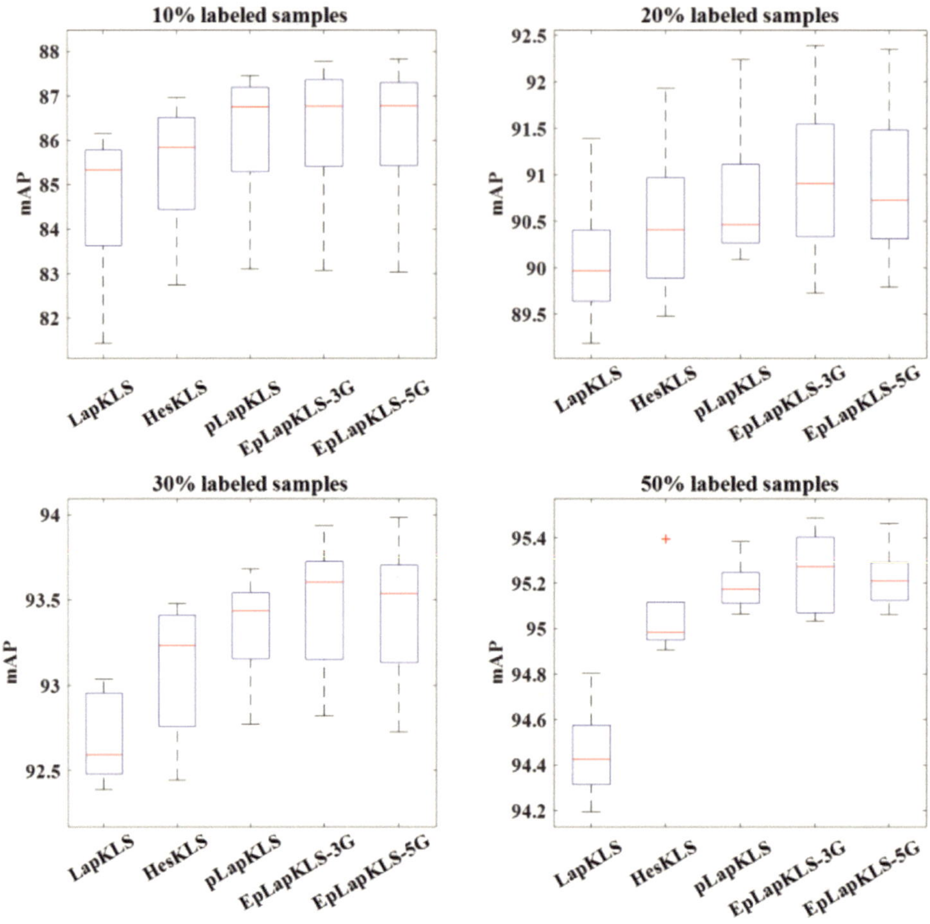

Figure 21. mAP performance of different algorithms on KLS method.

Here, an alternating optimization procedure is utilized to minimize f^*.

We compare EpLapR with other local structure preserving algorithms including LapR, HesR and pLapRon UC-Merced data set. We apply the support vector machines and kernel least squares for remote sensing image classification.

In the experiments, we apply the parameter setting as the same as pLapR, and the experiment of pLapR is conducted with $p = 2.8$. For EpLapR, we created two graph p-Laplacian sets. For the first set (EpLapR-3G), we choose $p = \{2.5, 2.7, 2.8\}$, which led to 3 graphs. For another one (EpLapR-5G), with 5 graphs where $p = \{2.4, 2.5, 2.6, 2.7, 2.8\}$.

We compare our proposed EpLapR with the representative LapR, HesR and pLapR. **Figures 21** and **22** demonstrate the mAP results of different algorithms on KLS methods and SVM methods, respectively. We can see that, in most cases, the EpLapR outperforms LapR, HesR and pLapR, which shows the advantages of EpLapR in local structure of preserving.

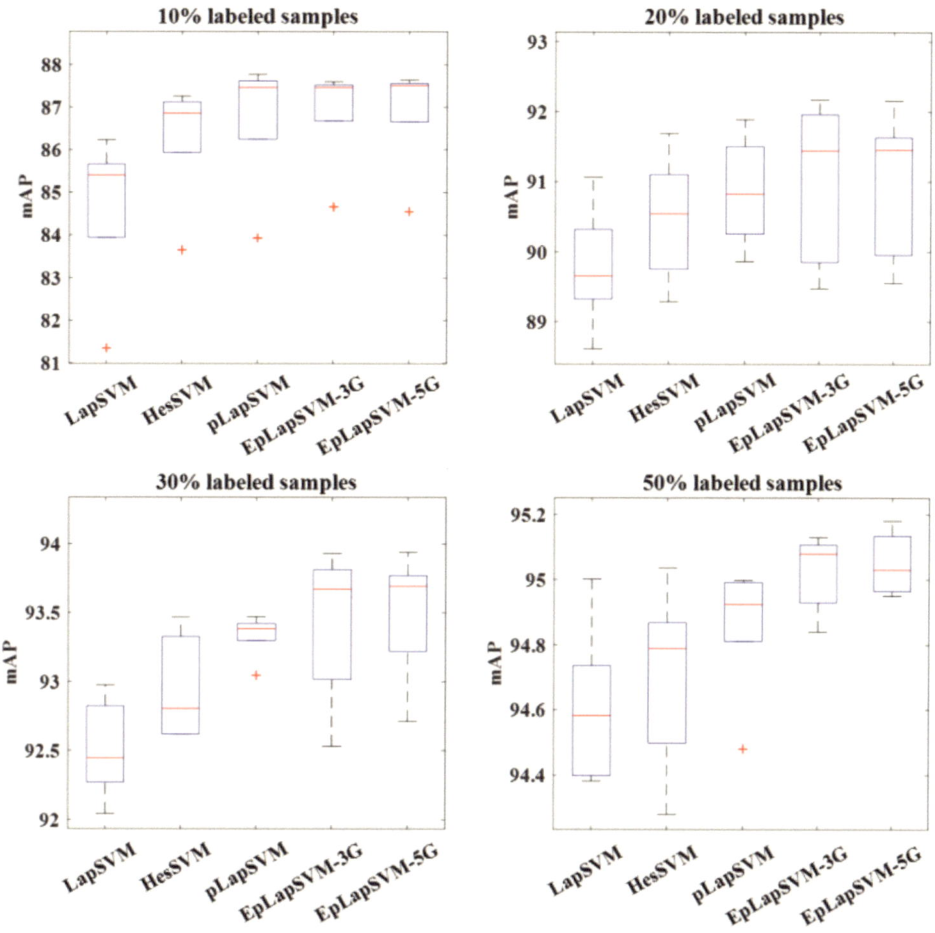

Figure 22. mAP performance of different algorithms on SVM method.

6. Conclusions

In this chapter, we show the LapR, HesR, pLapR and present several extensions based on the manifold regularization framework. We propose a local structure preserving method that effectively integrates manifold regularization and pairwise constraints. We develop an efficient approximation algorithm of graph p-Laplacian and propose p-Laplacian regularization to preserve the local geometry. Considering the hypergraph contains more local grouping information in comparison to simple graph, we propose hypergraph p-Laplacian regularization to preserve the geometry of the probability distribution. In practical application of p-Laplacian regularization model, it is difficult to determine the optimal graph p-Lapalcian because the parameter p usually chose by cross validation method which lacks the ability to approximate the optimal solution. Therefore, we propose an ensemble p-Laplacian regularization to better approximate the geometry of the data distribution.

7. Expectations

In the general image recognition, images are naturally represented by multi-view features, such as color, shape and texture. Each view of a feature summarizes a specific characteristic of the image, and features for different views are complementary to one another. Therefore, in the future work, we will study the multi-view p-Laplacian regularization to effectively explore the complementary properties of different features from different views. Meanwhile, we will try to combine the p-Laplacian learning with the deep learning to get a more effective p-Laplacian learning algorithm.

Author details

Xueqi Ma and Weifeng Liu*

*Address all correspondence to: liuwf@upc.edu.cn

China University of Petroleum (East China), Qingdao, China

References

[1] Kruskal JB. Multidimensional scaling by optimizing goodness of fit to a nonmetric hypothesis. Psychometrika. 1964;**29**(1):1-27. DOI: 10.1007/bf02289565

[2] Borg I, Groenen P. Modern multidimensional scaling: Theory and applications. Journal of Educational Measurement. 2003;**40**(3):277-280

[3] Bishop CM, Svensén M, Williams CKI. GTM: The generative topographic mapping. Neural Computation. 1998;**10**(1):215-234. DOI: 10.1162/089976698300017953

[4] Roweis ST, Saul LK. Nonlinear dimensionality reduction by locally linear embedding. Science. 2000;**290**(5500):2323-2326. DOI: 10.1126/science.290.5500.2323

[5] Bernstein M, Silva VD, Langford JC, et al. Graph approximations to geodesics on embedded manifolds. Stanford University. 2000;**24**(9):153-158

[6] Belkin M, Niyogi P. Laplacian eigenmaps and spectral techniques for embedding and clustering. Advances in Neural Information Processing Systems. 2001;**14**(6):585-591

[7] Donoho DL, Grimes C. Hessian eigenmaps: Locally linear embedding techniques for high-dimensional data. Proceedings of the National Academy of Sciences of the United States of America. 2003;**100**(10):5591. DOI: 10.1073/pnas.1031596100

[8] Zhang Z, Zha H. Principal manifolds and nonlinear dimensionality reduction via tangent space alignment. Society for Industrial and Applied Mathematics. 2005. DOI: 10.1137/s1064827502419154

[9] Wahba G. Spline models for observational data. Society for Industrial and Applied Mathematics. 1990. DOI: 10.1137/1.9781611970128

[10] Smola AJ, Bartlett P, Schölkopf B, et al. Regularization networks and support vector machines. Advances in Computational Mathematics. 2000;**13**(1):1-50

[11] Belkin M, Niyogi P, Sindhwani V. Manifold regularization: A geometric framework for learning from labeled and unlabeled examples. JMLR. 2006

[12] Geng B, Tao D, Xu C, et al. Ensemble manifold regularization. IEEE Transactions on Pattern Analysis and Machine Intelligence. 2012;**34**(6):1227-1233. DOI: 10.1109/TPAMI.2012.57

[13] Liu W, Liu H, Tao D, Wang Y, Lu K. Manifold regularized kernel logistic regression for web image annotation. Neurocomputing. Elsevier BV. 2016 Jan;**172**:3-8. DOI: 10.1016/j.neucom.2014.06.096

[14] Luo Y, Tao D, Geng B, Xu C, Maybank SJ. Manifold regularized multitask learning for semi-supervised multilabel image classification. IEEE Transactions on Image Processing. 2013;**22**(2):523-536. DOI: 10.1109/tip.2012.2218825

[15] Ma X, Tao D, Liu W. Effective human action recognition by combining manifold regularization and pairwise constraints. Multimedia Tools and Applications. 2017. DOI: 10.1007/s11042-017-5172-1

[16] Hu W, Cheung G, Li X, Au OC. Graph-based joint denoising and super-resolution of generalized piecewise smooth images. In: 2014 IEEE International Conference on Image Processing (ICIP). IEEE; 2014. DOI: 10.1109/icip.2014.7025412

[17] Kim KI, Steinke F, Hein M. Semi-supervised regression using hessian energy with an application to semi-supervised dimensionality reduction. Advances in Neural Information Processing Systems. 2009:979-987

[18] Liu W, Tao D, Cheng J, Tang Y. Multiview hessian discriminative sparse coding for image annotation. Computer Vision and Image Understanding. 2014;**118**:50-60. DOI: 10.1016/j.cviu.2013.03.007

[19] Liu W, Tao D. Multiview hessian regularization for image annotation. IEEE Transactions on Image Processing. 2013;**22**(7):2676-2687. DOI: 10.1109/tip.2013.2255302

[20] Liu X, Shi J, Wang C. Hessian regularization based non-negative matrix factorization for gene expression data clustering. In: Engineering in Medicine and Biology Society (EMBC), 2015 37th Annual International Conference of the IEEE. IEEE; 2015. pp. 4130-4133. DOI: 10.1109/embc.2015.7319303

[21] Zhu J, Shi J. Hessian regularization based semi-supervised dimensionality reduction for neuroimaging data of Alzheimer's disease. In: Biomedical Imaging (ISBI), 2014 IEEE 11th International Symposium on. IEEE; 2014. pp. 165-168. DOI: 10.1109/ISBI.2014.6867835

[22] Liu W, Yang X, Tao D, Cheng J, Tang Y. Multiview dimension reduction via hessian multiset canonical correlations. Information Fusion. 2018;**41**:119-128. DOI: 10.1016/j.inffus.2017.09.001

[23] Liu W, Liu H, Tao D. Hessian regularization by patch alignment framework. Neurocomputing. 2016;**204**:183-188. DOI: 10.1016/j.neucom.2015.07.152

[24] Liu W, Zhang L, Tao D, Cheng J. Support vector machine active learning by hessian regularization. Journal of Visual Communication and Image Representation. 2017;**49**:47-56. DOI: 10.1016/j.jvcir.2017.08.001

[25] Liu W, Li Y, Lin X, Tao D, Wang Y. Hessian-regularized co-training for social activity recognition. Chen K, editor. PLoS One. 2014;**9**(9):e108474. DOI: 10.1371/journal.pone.0108474

[26] Tao D, Jin L, Liu W, et al. Hessian regularized support vector machines for mobile image annotation on the cloud. IEEE Transactions on Multimedia. 2013;**15**(4):833-844. DOI: 10.1109/tmm.2013.2238909

[27] Liu W, Ma T, Tao D, You J. HSAE: A hessian regularized sparse auto-encoders. Neurocomputing. 2016;**187**:59-65. DOI: 10.1016/j.neucom.2015.07.119

[28] Liu W, Liu H, Tao D, Wang Y, Lu K. Multiview hessian regularized logistic regression for action recognition. Signal Processing. 2015;**110**:101-107. DOI: 10.1016/j.sigpro.2014.08.002

[29] Amghibech S. Eigenvalues of the discrete p-Laplacian for graphs. Ars Combinatoria-Waterloo then Winnipeg. 2003;**67**:283-302

[30] Bouchala J. Resonance problems for p-Laplacian. Mathematics and Computers in Simulation. 2003;**61**(3–6):599-604. DOI: 10.1016/s0378-4754(02)00139-8

[31] Bühler T, Hein M. Spectral clustering based on the graph p-Laplacian. In: Proceedings of the 26th Annual International Conference on Machine Learning (ICML'09). ACM Press; 2009. DOI: 10.1145/1553374.1553385

[32] Luo D, Huang H, Ding C, Nie F. On the eigenvectors of p-Laplacian. Machine Learning. . Springer Nature. 2010;**81**(1):37-51. DOI: 10.1007/s10994-010-5201-z

[33] Liu W, Zha Z-J, Wang Y, Lu K, Tao D. p-Laplacian regularized sparse coding for human activity recognition. IEEE Transactions on Industrial Electronics. 2016:1-1. DOI: 10.1109/tie.2016.2552147

[34] Wagstaff K, Cardie C. Clustering with instance-level constraints. International Conference on Machine Learning. 2000:1103-1110

[35] Davidson I, Basu S. A survey of clustering with instance level constraints. ACM Transactions on Knowledge Discovery from Data. 2007:1-41

[36] Fu Z, Lu Z, Ip HHS, Lu H, Wang Y. Local similarity learning for pairwise constraint propagation. Multimedia Tools and Applications. 2014;**74**(11):3739-3758. DOI: 10.1007/s11042-013-1796-y

[37] Bar-Hillel A, Hertz T, Shental N, et al. Learning a mahalanobis metric from equivalence constraints. Journal of Machine Learning Research. 2005;**6**(6):937-965

[38] Mignon A, Jurie F. PCCA: A new approach for distance learning from sparse pairwise constraints. In: 2012 IEEE Conference on Computer Vision and Pattern Recognition. IEEE; 2012. DOI: 10.1109/cvpr.2012.6247987

[39] Zhang D, Chen S, Zhou Z-H. Constraint score: A new filter method for feature selection with pairwise constraints. Pattern Recognition. 2008;**41**(5):1440-1451. DOI: 10.1016/j.patcog.2007.10.009

[40] Zhang D, Zhou ZH, Chen S. Semi-supervised dimensionality reduction. In: Siam International Conference on Data Mining, April 26–28, 2007, Minneapolis, Minnesota, USA. DBLP; 2007. pp. 11-393

[41] Cevikalp H, Verbeek JJ, Jurie F, et al. Semi-supervised dimensionality reduction using pairwise equivalence constraints. International Journal of System Dynamics Applications. 2008;**1**(3):489-496. DOI: 10.5220/0001070304890496

[42] Ding S, Jia H, Zhang L, Jin F. Research of semi-supervised spectral clustering algorithm based on pairwise constraints. Neural Computing and Applications. 2012;**24**(1):211-219. DOI: 10.1007 /s00521-012-1207-8

[43] Kalakech M, Biela P, Macaire L, Hamad D. Constraint scores for semi-supervised feature selection: A comparative study. Pattern Recognition Letters. 2011 Apr;**32**(5):656-665. DOI: 10.1016/j.patrec.2010.12.014

[44] Luo Y, Wen Y, Tao D. On combining side information and unlabeled data for heterogeneous multi-task metric learning. In: International Joint Conference on Artificial Intelligence. 2016. pp. 1809-1815

[45] Agarwal S, Lim J, Zelnik-Manor L, Perona P, Kriegman D, Belongie S. Beyond pairwise clustering. In: IEEE Computer Society Conference on Computer Vision and Pattern Recognition (CVPR'05). IEEE; 2005. DOI: 10.1109/cvpr.2005.89

[46] Ji R, Gao Y, Hong R, Liu Q, Tao D, Li X. Spectral-spatial constraint Hyperspectral image classification. IEEE Transactions on Geoscience and Remote Sensing. 2014 Mar;**52**(3):1811-1824. DOI: 10.1109/tgrs.2013.2255297

[47] Yu J, Rui Y, Tao D. Click prediction for web image reranking using multimodal sparse coding. IEEE Transactions on Image Processing. 2014;**23**(5):2019-2032. DOI: 10.1109/tip.2014.2311377

[48] Huang Y, Liu Q, Metaxas D. Video object segmentation by hypergraph cut. In: 2009 IEEE Conference on Computer Vision and Pattern Recognition. IEEE; 2009. DOI: 10.1109/cvprw.2009.5206795

[49] Zien JY, Schlag MDF, Chan PK. Multilevel spectral hypergraph partitioning with arbitrary vertex sizes. Computer-Aided Design of Integrated Circuits and Systems, IEEE Transactions on, 1997;**18**(9):1389-1399. DOI:10.1109/iccad.1996.569592

[50] Rodríguez JA. On the Laplacian spectrum and walk-regular hypergraphs. Linear and Multilinear Algebra. 2003;**51**(3):1-1. DOI: 10.1080/03081080306587

[51] Bolla M. Spectra, Euclidean representations and clusterings of hypergraphs. Discrete Mathematics. 1993;**117**(1–3):19-39. DOI: 10.1016/0012-365x(93)90322-k

[52] Zhou D, Huang J. Learning with hypergraphs: Clustering, classification, and embedding. In: International Conference on Neural Information Processing Systems. MIT Press; 2006. pp. 1601-1608

[53] Guo Y, Tao D, Liu W, Cheng J. Multiview Cauchy estimator feature embedding for depth and inertial sensor-based human action recognition. IEEE Transactions on Systems, Man, and Cybernetics: Systems. 2017;**47**(4):617-627. DOI: 10.1109/tsmc.2016.2617465

[54] Everingham M, Van Gool L, Williams CKI, Winn J, Zisserman A. The pascal visual object classes (VOC) challenge. International Journal of Computer Vision. 2009;**88**(2):303-338. DOI: 10.1007/s11263-009-0275-4

[55] Liu W, Ma X, Zhou Y, Tao D, Cheng J. P-Laplacian regularization for scene recognition. IEEE Transactions on Cybernetics. 2018. DOI: 10.1109/TCYB.2018.2833843

[56] Quattoni A, Torralba A. Recognizing indoor scenes. In: 2009 IEEE Conference on Computer Vision and Pattern Recognition. IEEE; 2009. DOI: 10.1109/cvprw.2009.5206537

[57] Lazebnik S, Schmid C, Ponce J. Beyond bags of features: Spatial pyramid matching for recognizing natural scene categories. In: IEEE Computer Society Conference on Computer Vision and Pattern Recognition (CVPR'06). IEEE; 2006. DOI: 10.1109/cvpr.2006.68

[58] Zhou D, Schölkopf B. Regularization on discrete spaces. In: Joint Pattern Recognition Symposium. Berlin, Heidelberg: Springer; 2005. pp. 361-368

[59] Takeuchi H. The spectrum of the p-Laplacian and p-harmonic morphisms on graphs. Illinois Journal of Mathematics. 2003;**47**(2003):939-955

[60] Yang Y, Newsam S. Bag-of-visual-words and spatial extensions for land-use classification. In: Proceedings of the 18th SIGSPATIAL International Conference on Advances in Geographic Information Systems - GIS '10. 2010. DOI: 10.1145/1869790.1869829

[61] Simonyan K, Zisserman A. Very deep convolutional networks for large-scale image recognition. Computer Science. 2014

[62] Huang Y, Liu Q, Zhang S, Metaxas DN. Image retrieval via probabilistic hypergraph ranking. In: 2010 IEEE Computer Society Conference on Computer Vision and Pattern Recognition. IEEE; 2010. DOI: 10.1109/cvpr.2010.5540012

[63] Wang M, Hua XS, Hong R, Tang J, et al. Unified video annotation via multigraph learning. IEEE Transactions on Circuits and Systems for Video Technology. 2009;**19**(5):733-746. DOI: 10.1109/tcsvt.2009.2017400

Section 2

Applications

Chapter 4

Manifold Learning in Medical Imaging

Samuel Kadoury

Additional information is available at the end of the chapter

http://dx.doi.org/10.5772/intechopen.79989

Abstract

Manifold learning theory has seen a surge of interest in the modeling of large and extensive datasets in medical imaging since they capture the essence of data in a way that fundamentally outperforms linear methodologies, the purpose of which is to essentially describe things that are flat. This problematic is particularly relevant with medical imaging data, where linear techniques are frequently unsuitable for capturing variations in anatomical structures. In many cases, there is enough structure in the data (CT, MRI, ultrasound) so a lower dimensional object can describe the degrees of freedom, such as in a manifold structure. Still, complex, multivariate distributions tend to demonstrate highly variable structural topologies that are impossible to capture with a single manifold learning algorithm. This chapter will present recent techniques developed in manifold theory for medical imaging analysis, to allow for statistical organ shape modeling, image segmentation and registration from the concept of navigation of manifolds, classification, as well as disease prediction models based on discriminant manifolds. We will present the theoretical basis of these works, with illustrative results on their applications from various organs and pathologies, including neurodegenerative diseases and spinal deformities.

Keywords: manifold learning, medical imaging, discriminant manifolds, piecewise geodesic regression, spine deformities, neurodegenerative diseases, shape modeling

1. Introduction

Learning on large medical imaging datasets is an emerging discipline driven from the availability of vast amounts of raw data in many of today's biomedical studies. However, challenges such as unbalanced data distributions, complex multivariate data and highly variable structural topologies demonstrated by real-world samples makes it much more difficult to efficiently learn the associated representation. An important goal of scientific data analysis in medicine, particularly in neurosciences or oncology, is to understand the behavior of biological process or physiological/morphological alterations. This introduces the need to synthesize large amounts of multivariate data in a robust manner and raises the fundamental question of data reduction: how to discover meaningful representations from unstructured high-dimensional medical images.

Several approaches have attempted to understand how dimension reduction and regression establishes the relationship in subspaces and finally determine statistics on manifolds that optimally describe the relationships between the samples [1]. However, certain assumptions based on the representation of shapes and images using smooth manifolds are made in most cases, which frequently will not be adequate in the presence of medical imaging data and often perturbed by nuisance articulations, clutter or varying contrast.

High-dimensional classification methods have shown promise to measure subtle and spatially complex imaging patterns that have diagnostic value [2, 3]. Defining statistics on a manifold is not a straightforward process when simple statistics cannot be directly applied to general manifolds [4]. But while Euclidean estimators have been used for vector spaces, none have been adapted for multimodal data lying in different spaces. Still, there has been interest in the characterization of data in a Riemann space [5, 6]. Unfortunately, manifold-valued metrics based on the centrality theory or the geometric median [7] often lacks robustness to outliers.

A related topic lies in dimensionally reduced growth trajectories of various anatomical sites which have been investigated in neurodevelopment of newborns for example, based on geodesic shape regression to compute the diffeomorphisms with image time series of a population [8]. These regression models were also used to estimate spatiotemporal evolution of the cerebral cortex [9]. The concept of parallel transport curves in the tangent space from low-dimensional manifolds proposed by Schiratti et al. [10] was used to analyze shape morphology [11] and adapted for radiotherapy response [12]. Regression models were proposed for both cortical and subcortical structures, with 4D varifold-based learning framework with local topography shape morphing being proposed by Rekik et al. [13].

This chapter presents several manifold learning methodologies designed to address challenges encountered in medical imaging. In Section 2, we present an articulated shape inference model from nonlinear embeddings, expressing the global and local shape variations of the spine and vertebrae composing it, introduced in [14]. We then present in Section 3 a probabilistic model from discriminant manifolds to classify the neurodegenerative stage of Alzheimer's disease. Finally, a piecewise-geodesic transport curve in the tangent space from low-dimensional manifolds designed for the prediction of correction in spinal surgeries is shown in Section 4, introducing a time-warping function controlling the rate of shape evolution. We conclude this article in Section 5.

2. Shape inference through navigating manifolds

Statistical models of shape variability have been successful in addressing fundamental vision tasks such as segmentation and registration in medical imaging. However, the high dimensionality and complex nonlinear underlying structure unfortunately makes the commonly used linear statistics inapplicable for anatomical structures. Manifold learning approaches map high-dimensional observation data that are presumed to lie on a nonlinear manifold, onto a single global coordinate system of lower dimensionality.

Inferring a model from the underlying manifold is not a novel concept but far from being trivial. In this section, we model both global statistics of the articulated model and local shape variations of vertebrae based on local measures in manifold space. We describe a spine inference/segmentation method from CT and MR images, where the model representation is optimized through a Markov Random Field (MRF) graph, balancing prior distribution with image data.

2.1. Data representation

Our spine model $\mathbf{S} = \{\mathbf{s}_1, \ldots, \mathbf{s}_L\}$ consists of an interconnection of L vertebrae. For each vertebra $\mathbf{s}_i$, we recover a triangular mesh with vertices $\left\{\mathbf{v}_j^i | j = 1, \ldots, V\right\}$, where the jth vertex corresponds to approximately the same location from one shape to another and V the number of vertices. Additionally, every $\mathbf{s}_i$ is annotated with landmarks on each model to rigidly register each object to its upper neighbor. Hence, an articulated deformable model (ADM) is represented by a vector of local intervertebral rigid transformations $A = [T_1, T_2, \ldots, T_L]$. To perform global shape modeling of $\mathbf{S}$, we convert A to an absolute representation $\mathbf{A}_{\text{abs}} = [T_1, T_1 \circ T_2, \ldots, T_1 \circ T_2 \circ \ldots \circ T_L]$ using recursive compositions. The transformations are expressed in the local coordinate system (LCS) of the lower vertebra. Center of transformation is the intersection of all three vertebral axes, following anteroposterior, cranial-caudal and left-right directions. Rigid transformations described here are the combination of a rotation matrix R, a translation t and scaling s. We formulate the rigid transformation $T = \{s, R, t\}$ of a triangular mesh model as $y = sRx + t$ where $x, y, t \in \Re^3$.

2.2. Manifold embedding

For nonlinear embeddings, we rely on the absolute vector representation $\mathbf{A}_{\text{abs}}$ as given previously. Let us now consider N articulated shape models expressed by the feature vectors $\mathbf{A}_{\text{abs}}^i$, of dimensionality D. The aim is to create a low-dimensional manifold consisting of N points $\mathbf{Y}_i$, $\mathbf{Y}_i \in \Re^d$, $i \in [1, N]$ where $d \ll D$ based on [15]. In such a framework, if an adequate number of data points is available, then the underlying manifold $\mathcal{M}$ is considered to be "well-sampled." Therefore, it can represent the underlying population structure. In the sub-cluster corresponding to a pathological population, each point of the training set and its neighbors would lie within a locally linear patch as illustrated in **Figure 1**.

The main limitation of embedding algorithms is the assumption of Euclidean metrics in the ambient space to evaluate similarity between sample points. Thus, a metric in the space of articulated structures is defined so that it accommodates for anatomical spine variability and adopts the intrinsic nature of the Riemannian manifold geometry allowing us to discern between articulated shape deformations in a topological invariant framework. For each point, the K closest neighbors are selected using a distortion metric which is particularly suited for geodesics. The metric $d_{\text{M}}\left(\mathbf{A}_{\text{abs}}^i, \mathbf{A}_{\text{abs}}^j\right)$ estimates the distance of articulated models i, j where

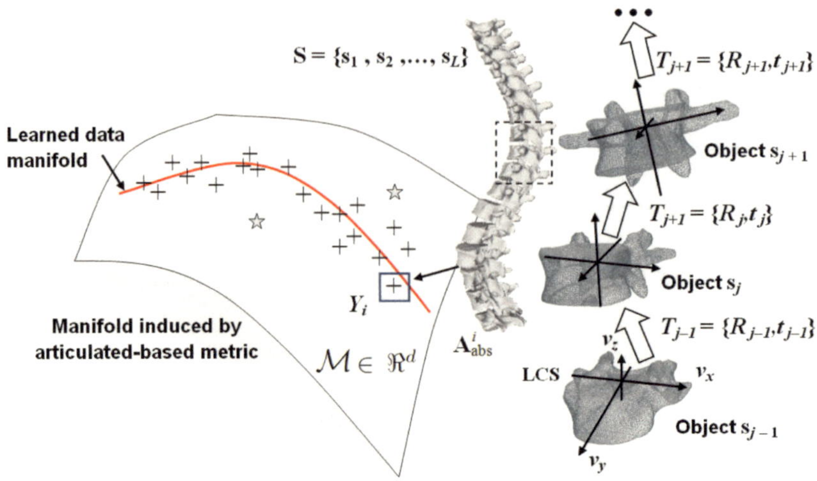

Figure 1. Representation of intervertebral transformations in manifold space.

$\mathbf{A}^i_{\text{abs}}$. The distance measure for absolute representations can therefore be expressed as a sum of articulation deviations

$$d_M\left(\mathbf{A}^i_{\text{abs}}, \mathbf{A}^j_{\text{abs}}\right) = \sum_{k=1}^{L} d_M\left(T^i_k, T^j_k\right) = \sum_{k=1}^{L} \|\mathbf{t}^i_k - \mathbf{t}^j_k\| + \sum_{k=1}^{L} d_G\left(\mathbf{R}^i_k, \mathbf{R}^j_k\right). \tag{1}$$

While for the translation, the L_2 norm is chosen, geodesical distances are used between rotation neighborhoods. This is expressed as $d_G\left(\mathbf{R}^i_k, \mathbf{R}^j_k\right) = \|\log\left(\left(\mathbf{R}^i_k\right)^{-1}\mathbf{R}^j_k\right)\|_F$ where the log map is used to map a point in the manifold to the tangent plane.

Afterwards, the manifold reconstruction weights are estimated by assuming the local geometry of the patches can be described by linear coefficients that permit the reconstruction of every model point from its neighbors. In order to determine the value of the weights, the reconstruction errors are measured using the following objective function:

$$\varepsilon(\mathbf{W}) = \sum_{i=1}^{N} \left\| \mathbf{A}^i_{\text{abs}} - \sum_{j=1}^{K} W_{ij}\mathbf{A}^j_{\text{abs}} \right\|^2 \tag{2}$$

$$\text{subject to} \begin{cases} W_{ij} = 0 \text{ if } \mathbf{A}^i_{\text{abs}} \text{ not neighbor } \mathbf{A}^j_{\text{abs}} \\ \sum_j \mathbf{W}_{ij} = 1 \text{ for every } i. \end{cases} \tag{3}$$

Thus, $\varepsilon(\mathbf{W})$ sums the squared distances between all data points and their corresponding reconstructed points. The weights $\mathbf{W}_{ij}$ represent the importance of the jth data point to the reconstruction of the ith element.

The algorithm maps each high-dimensional $\mathbf{A}^i_{\text{abs}}$ to a low-dimensional $\mathbf{Y}_i$. These internal coordinates are found with a cost function minimizing the reconstruction error:

$$\begin{aligned}\Phi(\mathbf{Y}) &= \sum_{i=1}^{N} \left\| \mathbf{Y}_i - \sum_{j=1}^{K} \mathbf{W}_{ij}\mathbf{Y}_j \right\|^2 \\ &= \sum_{i=1}^{N}\sum_{j=1}^{N} \mathbf{M}_{ij}\mathbf{Y}_i^T\mathbf{Y}_j\end{aligned} \tag{4}$$

with $\mathbf{M}$ as a sparse and symmetric $N \times N$ matrix enclosing the reconstruction weights $\mathbf{W}_{ij}$ such that $\mathbf{M} = (\mathbf{I} - \mathbf{W})^T(\mathbf{I} - \mathbf{W})$, and $\mathbf{Y}$ spanning the $\mathbf{Y}_i$'s. The optimal embedding, up to a global rotation, is obtained from the bottom $d + 1$ eigenvectors of $\mathbf{M}$ and helps to minimize the cost function $\Phi(\mathbf{Y})$ as a simple eigenvalue problem. The d eigenvectors form the d embedding coordinates. The coordinates $\mathbf{Y}_i$ can be translated by a constant displacement without affecting the overall cost $\Phi(\mathbf{Y})$. The eigenvector corresponding to the smallest eigenvalue corresponds to the mean value of the embedded data $\mathbf{Y}^0 = \{\mathbf{y}_1, \dots, \mathbf{y}_d\}, \mathbf{y}_i = 0, \forall i$. This can be discarded with $\sum \mathbf{Y}_i = 0$ to obtain an embedding centered at the origin. Hence, a new ADM can be inferred in the embedded d-space as a low-dimensional point $\mathbf{Y}^{\text{new}}$ by finding its optimal manifold coordinates $\mathbf{y}_i$.

To obtain the articulation vector for a new embedded point in the ambient space (image domain), one has to determine the representation in high-dimensional space based on its intrinsic coordinates. We first assume an explicit mapping $f : \mathcal{M} \to \Re^D$ from manifold space $\mathcal{M}$ to the ambient space $\Re^D$. The inverse mapping of $\mathbf{Y}_i$ is then performed by estimating the relationship between $\Re^D$ and $\mathcal{M}$ as a joint distribution, such there exists a smooth functional which belongs to a local neighborhood. Theoretically the manifold should follow the conditional expectation:

$$f(Y_i) \equiv E\left(\mathbf{A}^i_{\text{abs}} | \mathcal{M}(A_i) = \mathbf{Y}_i\right) = \int A_i \frac{p(\mathbf{Y}_i, A_i)}{p_{\mathcal{M}(A_i)}(\mathbf{Y}_i)} dD \tag{5}$$

which captures the overall trend of the data in D-space. Here, both $p_{\mathcal{M}(A_i)}(\mathbf{Y}_i)$ (marginal density of $\mathcal{M}(A_i)$) and $p(\mathbf{Y}_i, A_i)$ (joint density) are unknown. Based on the *Nadaraya-Watson* kernel regression [16], we replace densities by kernel functions as $p_{\mathcal{M}(A_i)}(\mathbf{Y}_i) = \frac{1}{K}\sum_{j \in \mathcal{N}(i)} G_h(\mathbf{Y}_i, \mathbf{Y}_j)$ and $p(\mathbf{Y}_i, A_i) = \frac{1}{K}\sum_{j \in \mathcal{N}(i)} G_h(\mathbf{Y}_i, \mathbf{Y}_j) G_g(A_i, A_j)$ [17]. The Gaussian regression kernels G require the neighbors $\mathbf{A}^j_{\text{abs}}$ of $j \in \mathcal{N}(i)$ to determine the bandwidths h, g so it includes all K data points ($\mathcal{N}(i)$ representing the neighborhood of i). Plugging these estimates in Eq.(5), this gives:

$$f_{\text{NW}}(\mathbf{Y}_i) = \int A_i \frac{\frac{1}{K}\sum_{j \in \mathcal{N}(i)} G_h(\mathbf{Y}_i, \mathbf{Y}_j) G_g(A_i, A_j)}{\frac{1}{K}\sum_{j \in \mathcal{N}(i)} G_h(\mathbf{Y}_i, \mathbf{Y}_j)} dD. \tag{6}$$

By assuming G is symmetric about the origin, we propose to integrate in the kernel regression estimator, the manifold-based distortion metric d_M which is particularly suited for geodesic metrics and articulated diffeomorphisms. This generalizes the expectation such that the observations $\mathbf{Y}$ are defined in manifold space $\mathcal{M}$:

$$f_{\mathrm{NW}}(\mathbf{Y}_i) = \underset{\mathbf{A}^i_{\mathrm{abs}}}{\mathrm{argmin}} \frac{\sum_{j\in\mathcal{N}(i)} G(\mathbf{Y}_i, \mathbf{Y}_j) d_M\left(\mathbf{A}^i_{\mathrm{abs}}, \mathbf{A}^j_{\mathrm{abs}}\right)}{\sum_{j\in\mathcal{N}(i)} G(\mathbf{Y}_i, \mathbf{Y}_j)} \tag{7}$$

which integrates the distance metric $d_M\left(\mathbf{A}^i_{\mathrm{abs}}, \mathbf{A}^j_{\mathrm{abs}}\right)$ defined in Eq. (1) and updates $f_{\mathrm{NW}}(\mathbf{Y}_i)$ using the closest neighbors of point $\mathbf{Y}_i$ in the manifold space. This constrains the regression to be valid for similar data points in its vicinity since locality around $\mathbf{Y}_i$ preserves locality in $\mathbf{A}^i_{\mathrm{abs}}$.

2.3. Optimization on manifold

Once an appropriate modeling of spine shape variations is determined with a manifold, a successful inference between the image and manifold must be accomplished. We describe here how a new model is generated. We search the optimal embedded manifold point $\mathbf{Y} = (\mathbf{y}_1, \ldots, \mathbf{y}_d)$ of the global spine model. Such a strategy offers an ideal compromise between the prior constraints, as well as the individual shape variations described by the weight vector $\mathbf{W} = (\mathbf{w}_1, \ldots, \mathbf{w}_n)$ in a localized sub-patch. The energy E of inferring the model $\mathbf{S}$ in the image $\mathcal{I}$ is a function of the set of displacement vectors Δ in the manifold space for global shape representation. This involves: (a) a data-related term expressing the image cost and (b) a global prior term measuring deformation between low-dimensional vectors with shape models. The third term represents (c) a higher-order term which is expressed by the reconstruction weights Ω for local vertebra modeling. The energy E can be expressed as the following combination of a global and local optimization:

$$E(\mathbf{S}^0, \mathcal{I}, \Delta, \Omega) = V(\mathbf{Y}^0 + \Delta, \mathcal{I}) + \alpha\, V(\mathbf{N}, \Delta) + \beta\, V(\mathbf{H}, \Delta, \Omega). \tag{8}$$

The global alignment of the model with the target image primarily drives the deformation of the model. The purpose is to estimate the set of articulations describing the global spine model by determining its optimal representation $\mathbf{Y}^0$ in the embedded space. This is performed by obtaining the global representation using the mapping in (7) so that: $f_{\mathrm{NW}}(Y_i + \Delta) = f_{\mathrm{NW}}(\{y_1 + \boldsymbol{\delta}_1, \ldots, y_d + \boldsymbol{\delta}_d\})$. This allows to optimize the model in manifold space coordinates while retrieving the articulations in $\mathcal{I}$. The global cost can be expressed as:

$$V(\mathbf{Y}^0 + \Delta, \mathcal{I}) = V(f_{\mathrm{NW}}(\{y_1 + \boldsymbol{\delta}_1, \ldots, y_d + \delta_d\}), \mathcal{I})). \tag{9}$$

The inverse transform allows to obtain $\mathbf{A}^i_{\mathrm{abs}} + \boldsymbol{D}$, with $\boldsymbol{D}$ as deformations in the image space. Since the transformations T_i are implicitly modeled in the absolute representation $\mathbf{A}^0_{\mathrm{abs}}$, we can formally consider the singleton image-related term as a summation of costs associated with each L vertebra of the model:

http://dx.doi.org/10.5772/intechopen.79989

$$V\left(\mathbf{A}^{0}_{\mathrm{abs}}+\boldsymbol{D},\mathcal{I}\right)=\sum_{i=1}^{L}V_{i}\left(\mathbf{s}_{i}{}^{*}\left(T_{i}^{0}+d_{i}\right),\mathcal{I}\right) \tag{10}$$

where $V_i(\mathbf{s},\mathcal{I})=\sum_{\mathbf{v}_i\in\mathbf{s}}\mathbf{n}_i^T(\mathbf{v}_i)\nabla\mathcal{I}(\mathbf{v}_i)$ minimizes the distance between mesh vertices of the inferred shape and gradient image $\mathcal{I}$ by a rigid transformation. Here, $\mathbf{n}_i$ is the normal pointing outwards and $\nabla\mathcal{I}(\mathbf{v}_i)$ the image gradient at $\mathbf{v}_i$.

The prior constraint for the rigid alignment are pairwise potentials between neighboring models y_i such that the difference in manifold coordinates is minimal with regards to a prior distribution of neighboring distances P:

$$\alpha V(\mathbf{N},\Delta)=\alpha\sum_{i\in G}\sum_{j\in\mathcal{N}(i)}V_{ij}\left(y_i^0+\boldsymbol{\delta}_i,y_j^0+\boldsymbol{\delta}_j,P\right). \tag{11}$$

This term represents the smoothness term of the global cost function to ensure that the deformation $\boldsymbol{\delta}_i$ applied to point coordinates are regular, with $V_{ij}=(0,1)$ a distance assigning function based on the distances to P.

One can integrate the global data and prior terms along with local shape terms parameterized as the higher-order cliques, by combining (9), (11):

$$\begin{aligned}E\left(\mathbf{S}^0,\mathcal{I},\Delta,\Omega\right)&=V\left(f_{\mathrm{NW}}\left(\{y_1+\boldsymbol{\delta}_1,\ldots,y_d+\boldsymbol{\delta}_{\mathbf{d}}\}\right),\mathcal{I}\right))\\&+\alpha\sum_{i\in G}\sum_{j\in\mathcal{N}(i)}V_{ij}\left(y_i^0+\boldsymbol{\delta}_i,y_j^0+\boldsymbol{\delta}_j\right)+\beta\sum_{c\in\mathcal{C}}V_c\left(\mathbf{w}_c^0+\omega_c\right).\end{aligned} \tag{12}$$

The optimization strategy of the resulting MRF (12) in the continuous domain is not a straightforward problem. The convexity of the solution domain is not guaranteed, while gradient-descent optimization approaches are prone to nonlinearity and local minimums. We seek to assign the optimal labels $\mathcal{L}^\Delta=\{l_1,\ldots,l_d\}$ and $\mathcal{L}^\Omega=\{l_1,\ldots,l_n\}$ which are associated to the quantized space Δ of displacements and local weight parameters Ω respectively. We consider that displacing the coordinates of point y_i^0 by $\boldsymbol{\delta}^{l_i}$ is equivalent to assigning label l_i to y_i^0. An incremental approach is adopted where in each iteration t we look for the set of labels that improves the current solution s.t. $y_i^t=y_i^0+\sum_t\boldsymbol{\delta}^{l_i t}$, which is a temporal minimization problem. Then (12) can be rewritten as:

$$\begin{aligned}E^t\left(\mathcal{L}^\Delta,\mathcal{L}^\Omega\right)&=V\left(f_{\mathrm{NW}}\left(\{y_1^{t-1},l_1^\Delta,\ldots,y_d^{t-1},l_d^\Delta\}\right),\mathcal{I}\right))\\&+\alpha\sum_{i\in G}\sum_{j\in\mathcal{N}(i)}V_{ij}\left(y_i^{t-1},y_j^{t-1},l_i^\Delta,l_j^\Delta\right)+\beta\sum_{c\in\mathcal{C}}V_c\left(\mathbf{w}_c^{t-1},l_c^\Omega\right).\end{aligned} \tag{13}$$

We solve the minimization of the higher-order cliques in (13) by transforming them into quadratic functions [18]. We apply the FastPD method [19] which solves the problem by formulating the duality theory in linear programming.

2.4. Results

Manifold learning. The manifold was built from a database containing 711 scoliotic spines demonstrating several types of deformities. Each spine model in the database was obtained from biplanar radiographic stereo-reconstructions. It is modeled with 12 thoracic and 5 lumbar vertebrae (17 in total), represented by 6 landmarks on each vertebra (4 pedicle extremities and 2 endplate center points) which were manually identified by an expert on the radiographic images. The resulting manifold is shown in **Figure 2**.

Adaptation of the articulated model was done on two different data sets. The first consisted of volumetric CT scans ($512 \times 512 \times 251$, resolution: 0.8×0.8 mm, thickness: $1 - 2$ mm) of the lumbar and main thoracic regions obtained from 21 different patients acquired for operative planning purposes. The MR dataset comprised multi-parametric volumetric data ($256 \times 256 \times 160$, resolution: 1.3×0.9 mm, thickness: 1 mm) of 8 patients acquired for diagnostic purposes. For this study, only the T1 sequence was selected for the experiments. All patients on both datasets (29 in total) had 12 thoracic and 5 lumbar vertebrae. Both CT and MR data were manually annotated with 3D landmarks by an expert in radiology, corresponding to left and right pedicle tips as well as midpoints of the vertebral body. Segmentation of the vertebrae from the CT and MR slices were also made by the same operator.

CT imaging experiments. We first evaluated the model accuracy in CT images by computing the correspondence of the inferred vertebral mesh models to the segmented target structures. As a preprocessing step, a rough thresholding was performed on the whole volume to filter out noise artifacts. The overall surface-to-surface comparison results between the inferred 3D

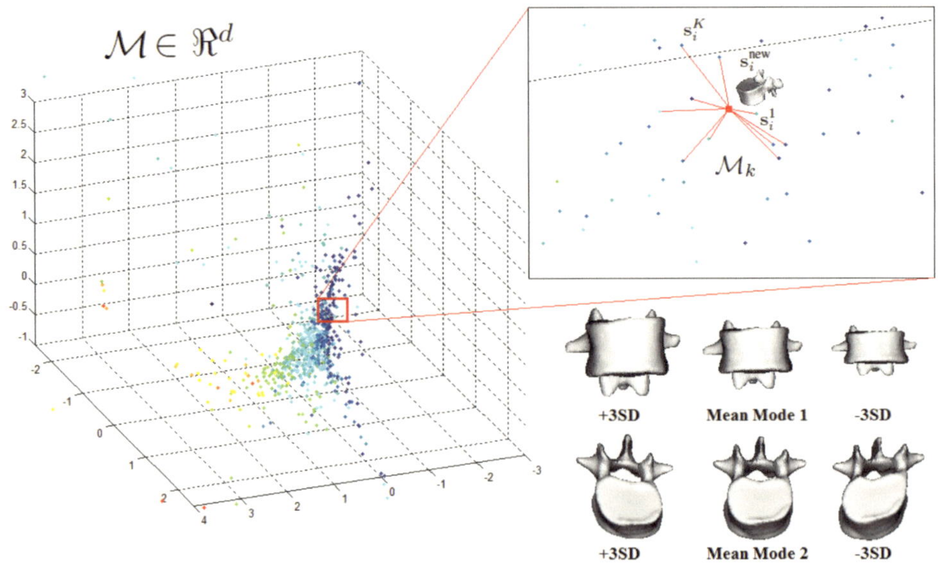

Figure 2. Low-dimensional manifold embedding of the spine dataset comprising 711 models exhibiting various types of deformities. The sub-domain was used to estimate both the global shape pose costs and individual shape instances based on local neighborhoods.

http://dx.doi.org/10.5772/intechopen.79989

vertebral models issued from the articulated model and from known segmentations were first calculated. The mean errors are 2.2 ± 1.5 mm (range: $0.6 - 5.4$ mm) for thoracic vertebra and 2.8 ± 1.9 mm (range: $0.7 - 8.1$ mm) for lumbar vertebra.

MR imaging experiments. For the experiments involving the segmentation of 3D spine models from MR images, the surface-to-surface comparison showed encouraging results (thoracic: 2.9 ± 1.8 mm, lumbar: 3.0 ± 1.9 mm) based on differences to ground-truth. As in the previous experiments with CT imaging, ground-truth data was generated by manually segmenting the structures models which were validated by an expert in radiology. As difficult as the CT inference is, the MR problem represent an even greater challenge as the image resolution is more limited and interslice spacing is increased compared to CT. Modeling of the statistical properties of the shape variations and global pose becomes even more important in this case, as it relies heavily in the nonlinear distribution of the patient morphology.

3. Probabilistic modeling of discriminant nonlinear manifolds in the identification of Alzheimer's

Neurodegenerative pathologies, such as Alzheimer's disease (AD), are linked with morphological and metabolic alterations which can be assessed from medical imaging and biological data. Recent advances in machine learning have helped to improve classification and prognosis rates, but lack a probabilistic framework to measure uncertainty in the data. In this section, we present a method to identify progressive mild cognitive impairment (MCI) and predict their conversion to AD from MRI and positron emitting tomography (PET) images. We show a discriminative probabilistic manifold embedding where locally linear mappings transform data points in low-dimensional space to corresponding points in high-dimensional space. A discriminant adjacency matrix is constructed to maximize the separation between different clinical groups, including MCI converters and nonconverters, while minimizing the distance in latent variables belonging to the same class.

3.1. Probabilistic model for discriminant manifolds

Manifold learning algorithms are based on the premise that data are often of artificially high dimension and can be embedded in a lower dimensional space. However the presence of outliers and multiclass information can on the other hand affect the discrimination and/or generalization ability of the manifold. We propose to learn the optimal separation between four classes (1) normal controls, (2) nonconverter MCI patients, (3) converter MCI patients and (4) AD patients, by using a discriminant graph-embedding. Here, n labeled points $\mathbb{Y} = \{(y_i, l_i)\}_{i=1}^{n}$ defined in $\mathbb{R}^D$ are generated from the underlying manifold $\mathcal{M}$, where l_i denotes the label (NC, cMCI, nMCI or AD). For the labeled data, there exists a low-dimensional (latent) representation of the high-dimensional samples such that $\mathbb{X} = \{(\mathbf{x}_i, l_i)\}_{i=1}^{n}$ defined in $\mathbb{R}^d$. We assume here that the mapping $\mathbf{M}_i \in \mathbb{R}^{D \times d}$ between high and low-dimensional spaces is locally linear, such that tangent spaces in local neighborhoods can be estimated with $\mathbf{y}_j - \mathbf{y}_i$ and

$\mathbf{x}_j - \mathbf{x}_i$, representing the pairwise differences between connected neighbors i, j. Therefore the relationship can be established as $\mathbf{y}_j - \mathbf{y}_i \approx \mathbf{M}_i(\mathbf{x}_j - \mathbf{x}_i)$.

In order to effectively discover the low-dimensional embedding, it is necessary to maintain the local structure of the data in the new embedding. The graph $G = (\boldsymbol{V}, \boldsymbol{W})$ is an undirected similarity graph, with a collection of nodes $\boldsymbol{V}$ connected by edges, and the symmetric matrix $\boldsymbol{W}$ with elements describing the relationships between the nodes. The diagonal matrix $\boldsymbol{D}$ and the Laplacian matrix $\boldsymbol{L}$ are defined as $\boldsymbol{L} = \boldsymbol{D} - \boldsymbol{W}$, with $\boldsymbol{D}(i, i) = \sum_{j \neq i} \boldsymbol{W}_{ij} \forall i$.

Using the theoretical framework from [20], we can determine a distribution of linear maps associated with the low-dimensional representation to describe the data likelihood for a specific model:

$$\log p(\mathbb{Y}|G) = \log \int\int p(\mathbb{Y}, \mathbf{M}, \mathbb{X}|G) d\mathbf{x} d\mathbf{M} \tag{14}$$

This joint distribution can be separated into three prior terms: the linear maps, latent variables and the likelihood of the high dimensional points $\mathbb{Y}$:

$$p(\mathbb{Y}, \mathbf{M}, \mathbb{X}|G) = p(\mathbb{Y}|\mathbf{M}, \mathbb{X}, G) p(\mathbf{M}|G) p(\mathbb{X}|G) \tag{15}$$

We now define the discriminant similarity graphs establishing neighborhood relationships, as well define each of the three prior terms included in the joint distribution.

Within and between similarity graphs: In our work, the geometrical structure of $\mathcal{M}$ can be modeled by building a within-class similarity graph $\boldsymbol{W}_w$ for feature vectors of same group and a between-class similarity graph $\boldsymbol{W}_b$, to separate features from all four classes. When constructing the discriminant locally linear latent variable embedding, elements are partitioned into $\boldsymbol{W}_w$ and $\boldsymbol{W}_b$ classes. The intrinsic graph G is first created by assigning edges only to samples of the same class (ex: nMCI). Each sample is therefore reconstructed only from feature vectors of the same clinical group. Local reconstruction coefficients are incorporated in the within-class similarity graph, such that $\boldsymbol{W}_w$ is defined as:

$$W_{w_{i,j}} = \begin{cases} 1 & \text{if } \mathbf{y}_i \in \mathcal{N}_w(\mathbf{y}_j) \text{ or } \mathbf{y}_j \in \mathcal{N}_w(\mathbf{y}_i) \\ 0, & \text{otherwise.} \end{cases} \tag{16}$$

with $\mathcal{N}_w$ containing neighbors of the same class. Conversely, $\boldsymbol{W}_b$ depicts the statistical properties to be avoided in the inference process. Distances between samples from different clinical groups are computed as:

$$W_{b_{i,j}} = \begin{cases} 1 & \text{if } \mathbf{y}_i \in \mathcal{N}_b(\mathbf{y}_j) \text{ or } \mathbf{y}_j \in \mathcal{N}_b(\mathbf{y}_i) \\ 0, & \text{otherwise} \end{cases} \tag{17}$$

with $\mathcal{N}_b$ containing neighbors having different class labels from the ith sample. The objective is to transform points to a new manifold $\mathcal{M}$ of dimensionality d, i.e., $\mathbf{y}_i \rightarrow \mathbf{x}_i$, by mapping connected samples from the same group in $\boldsymbol{W}_w$ as close as possible to the class cluster, while moving NC, nMCI, cMCI and AD samples of $\boldsymbol{W}_b$ as far away from one another.

Model components: The prior added on the latent variables $\mathbb{X}$ are located at the origin of the low-dimensional domain, while minimizing the Euclidean distance of neighboring points that are associated with the neighborhood of high-dimensional points and maximizing the distance between coordinates of different classes. In order to set the variables with an expected scale α and H representing the probability density function, the following log prior is defined:

$$\log p(\mathbb{X}|\boldsymbol{W},\alpha) = -\frac{1}{2}\sum_{i=1}^{n}\left(\alpha\|\mathbf{x}_i\| + \sum_{j=1}^{n} W_{w_{i,j}}\|\mathbf{y}_i - \mathbf{y}_j\|^2 - \sum_{j=1}^{n} W_{b_{i,j}}\|\mathbf{y}_i - \mathbf{y}_j\|^2\right) - \log H_{\mathbb{X}} \quad (18)$$

The prior added to the linear maps defines how the tangent planes described in low and high dimensional spaces are similar based on the Frobenius norm. This prior ensures smooth manifolds:

$$\log p(\mathbf{M}|\boldsymbol{W}) = -\frac{1}{2}\left(\left\|\sum_{i=1}^{n}\mathbf{x}_i\right\|_F^2 - \sum_{i=1}^{n}\sum_{j=1}^{n}\left(W_{w_{i,j}} - W_{b_{i,j}}\right)\|\mathbf{M}_i - \mathbf{M}_j\|_F^2\right) - \log H_{\mathbf{M}} \quad (19)$$

Finally, approximation errors from the linear mapping $\mathbf{M}_i$ between low and high-dimensional domains are penalized by including the following log likelihood:

$$\begin{aligned}\log p(\mathbb{Y}|\mathbb{X},\mathbf{W},\gamma) = \; & \|\sum_{i=1}^{n}\mathbf{y}_i\|^2 - \frac{1}{2}\sum_{i=1}^{n}\sum_{j=1}^{n} W_{w_{i,j}}\Delta(i,j)^{\mathrm{T}}\gamma\mathbf{I}\Delta(i,j) \\ & + \frac{1}{2}\sum_{i=1}^{n}\sum_{j=1}^{n} W_{b_{i,j}}\Delta(i,j)^{\mathrm{T}}\gamma\mathbf{I}\Delta(i,j) - \log H_{\mathbf{y}}\end{aligned} \quad (20)$$

with $\Delta(i,j)$ the difference in Euclidean distance between pairs of neighbors in high and low-dimensional space and γ the update parameter for the EM inference. Samples of $\mathbf{y}$ are drawn from a multivariate normal distribution.

3.2. Variational inference

The objective is to infer the low-dimensional coordinates and linear mapping function for the described model, as well as the intrinsic parameters of the model $\Phi = (\alpha, \gamma)$. This is achieved by maximizing the marginal likelihood of:

$$\log p(\mathbb{Y}|\mathbf{W},\Phi) = \iint \rho(\mathbf{M},\mathbb{X}) \log \frac{p(\mathbb{Y},\mathbf{M},\mathbb{X}|\mathbf{W},\Phi)}{\rho(\mathbf{M},\mathbb{X})} d\mathbf{x}d\mathbf{M}. \quad (21)$$

By assuming the posterior $\rho(\mathbf{M},\mathbb{X})$ can be factored in separate terms $\rho(\mathbf{M})$ and $\rho(\mathbb{X})$, a variational expectation maximization algorithm can be used to determine the model's parameters, which are initialized with Φ. The E-step updates the independent posteriors $\rho(\mathbb{X})$ and $\rho(\mathbf{M})$, while the parameters of Φ are updated in the M-step by maximizing Eq. (21).

The discriminant latent variable model can then be used to perform the mapping of new image feature vectors to the manifold. The variational EM algorithm described in the previous section can be used to transform a set of new input points $\mathbf{y}_q$ without changing the overall neighborhood graph structure, by finding the distribution of the local linear map $\mathbf{y}_q$ and it is low-dimensional coordinate using the E-step explained above. Once the manifold representation $\mathbf{x}_q$ is obtained, a cluster analysis finds the corresponding class in the manifold, yielding a prediction of the input feature vector $\mathbf{y}_q$.

3.3. Experiments

We used the Alzheimer's Disease Neuroimaging Initiative (ADNI) database with 1.5 or 3.0 T structural MR images (adni.loni.usc.edu) and FDG-PET images. For this study, 187 subjects with both MRI and PET images during a 24 month period were used to train the probabilistic manifold model, including 46 AD patients, 94 MCI patients, and 47 normal controls. During the follow-up period, 43 MCI subjects converted to AD and 56 remained stable. All groups are matched approximately by age (mean of 76.7 ± 5.4) and gender. Images were non-rigidly registered to a standard template, which was then segmented using FSL-FIRST automatic segmentation [21].

A 9-fold cross-validation was performed to assess the performance of the method. The optimal manifold dimensionality was set at $d = 8$, when the trend of the nonlinear residual reconstruction error stabilized for the entire training set. We evaluated the classification performance of the proposed method for discriminating between cMCI and nMCI patients, by training the model with MRI, PET and with MRI + PET biomarkers from the ROIs illustrated in **Figure 3**. **Figure 4** presents ROC curves obtained by the proposed and comparative methods such as SVM (nonlinear RBF kernel), LLE and LL-LVM [20]. The discriminative nature of the proposed framework clearly shows an improvement to standard learning approaches models which were trained using MRI only, PET only and combined multimodal features. It illustrates that increased accuracy (77.4%) can be achieved by combining MRI and PET features, showing the benefit of extracting complementary features from the dataset for prediction purposes. When comparing the performance of the proposed method to the other learning methods (SVM, LLE,

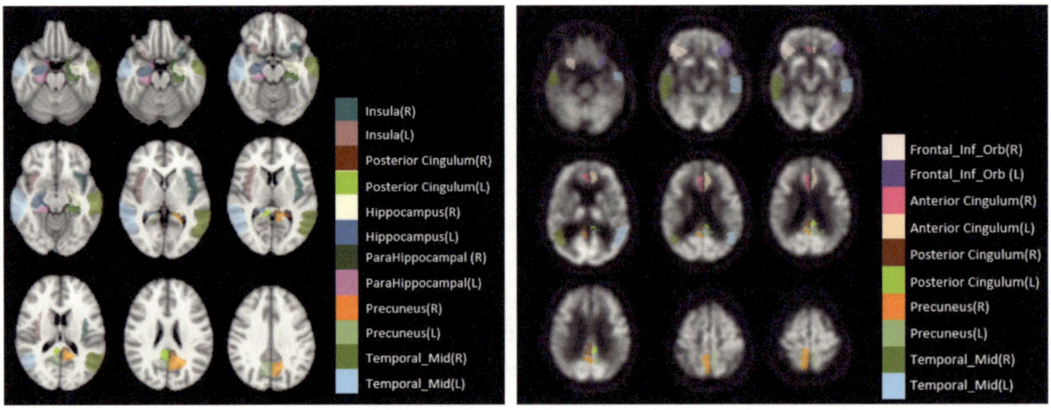

Figure 3. Selected FSL segmented brain regions for feature selection on (left) MRI and (right) PET images.

http://dx.doi.org/10.5772/intechopen.79989

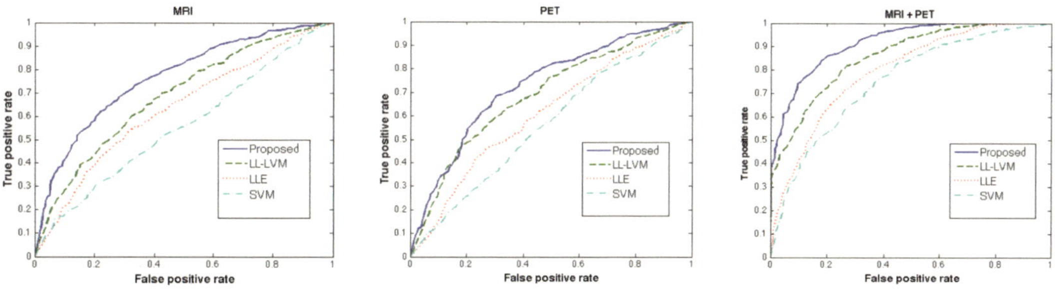

Figure 4. ROC curves comparing the SVM, LLE and LL-LVM with the proposed method for cMCI/nMCI prediction using MRI, PET and multimodality data.

LL-LVM), the probabilistic model integrating similarity graphs shows a statistically significant improvement ($p < 0.01$) to all three approaches based on paired t-test.

4. Spatiotemporal manifold prediction model for surgery prediction

In this final section, we present a statistical framework for predicting the surgical outcomes following spine surgery of adolescents with idiopathic scoliosis. A discriminant manifold is first constructed to maximize the separation between responsive and nonresponsive groups of patients. The model then uses subject-specific correction trajectories based on articulated transformations in order to map spine correction profiles to a group-average piecewise-geodesic path. Spine correction trajectories are described in a piecewise-geodesic fashion to account for varying times at follow-up exams, regressing the curve via a quadratic optimization process. To predict the evolution of correction, a baseline reconstruction is projected onto the manifold, from which a spatiotemporal regression model is built from parallel transport curves inferred from neighboring exemplars (**Figure 5**).

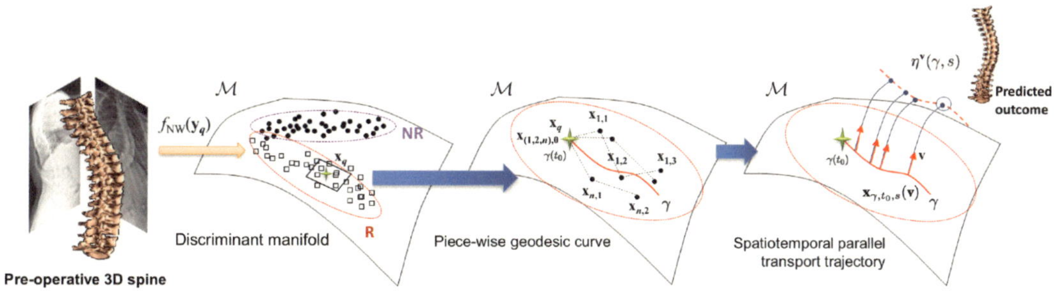

Figure 5. Proposed prediction framework for spine surgery outcomes. In the training phase, a dataset of spine models are embedded in a spatiotemporal manifold $\mathcal{M}$, into responsive (R) or nonresponsive (NR) groups. During testing, an unseen baseline 3D spine reconstruction $\mathbf{y}_q$ is projected on $\mathcal{M}$ using f_{NW} based on Nadaraya-Watson kernels. The closest samples to the projected point $\mathbf{x}$ are selected to regress the spatiotemporal curve γ used for predicting the correction due with surgery.

4.1. Discriminant embedding of spine models

We propose to embed a collection of nonresponsive (NR) and (2) responsive (R) patients to surgery which will offer a maximal separation between the classes, by using a discriminant graph-embedding. Here, n labeled points $\mathbb{Y} = \{(\mathbf{y}_i, l_i, t_i)\}_{i=1}^{n}$ defined in $\mathbb{R}^D$ are embedded in the low-dimensional manifold $\mathcal{M}$, where l_i describes the label (NR or R) and t_i defines the time of follow-up. We assume that for the sampled data, an underlying manifold of the high-dimensional data exists such that $\mathbb{X} = \{(\mathbf{x}_i, l_i, t_i)\}_{i=1}^{n}$ defined in $\mathbb{R}^d$. We rely on the assumption that a locally linear mapping $\mathbf{M}_i \in \mathbb{R}^{D \times d}$ exists, where local neighborhoods are defined as tangent planes estimated with $\mathbf{y}_j - \mathbf{y}_i$ and $\mathbf{x}_j - \mathbf{x}_i$, describing the paired distances between linked neighbors i, j. Hence, the relationship can be established as $\mathbf{y}_j - \mathbf{y}_i \approx \mathbf{M}_i(\mathbf{x}_j - \mathbf{x}_i)$.

Because the discriminant manifold structure in $\mathbb{R}^d$ requires to maintain the local structure of the underlying data, a undirected similarity graph $\mathcal{G} = (\boldsymbol{V}, \boldsymbol{W})$ is built, where each node $\boldsymbol{V}$ are connected to each other with edges that are weighted with the graph $\boldsymbol{W}$. The overall structure of $\mathcal{M}$ is therefore defined with $\boldsymbol{W}_w$ for feature vectors belonging to the same class and $\boldsymbol{W}_b$, which separate features from both classes. During the embedding of the discriminant locally linear latent manifold, data samples are divided between $\boldsymbol{W}_w$ and $\boldsymbol{W}_b$.

4.2. Piecewise-geodesic spatiotemporal manifold

Once sample points $\mathbf{x}_i$ are in manifold space, the objective is to regress a regular and smooth piecewise-geodesic curve $\gamma : [t_1, t_N]$ that accurately fits the embedded data describing the spatiotemporal correction following surgery within a 2 year period. For each sample data $\mathbf{x}_i$, the K closest individuals demonstrating similar baseline features are identified from the embedded data, creating neighborhoods $\mathcal{N}(\mathbf{x}_q)$ with measurements at different time points, thus creating a low-dimensional Riemannian manifold where data points $\mathbf{x}_{i,j}$, with i denoting a particular individual, j the time-point measurement and $j = 0$ the preoperative model. By assuming the manifold domain is complete and piecewise-geodesic curves are defined for each time trajectories, time-labeled data can be regressed continuously in $\mathbb{R}^D$, thereby creating smooth curves in time intervals described by samples in $\mathbb{R}^d$.

However, due to the fact the representation of the continuous curve is a variational problem of infinite dimensional space, the implementation follows a discretization process which is derived from the procedure in [22], such that:

$$
\begin{aligned}
E(\gamma) = & \frac{1}{K_d} \sum_{i=1}^{K_d} \sum_{j=0}^{t_N} w_i \| \gamma(t_{i,j}) - (\mathbf{x}_{i,j} - (\mathbf{x}_{i,0} - \mathbf{x}_q)) \|^2 \\
& + \frac{\lambda}{2} \sum_{i=1}^{K_d} \alpha_i \| v_i \|^2 + \frac{\mu}{2} \sum_{i=1}^{K_d} \beta_i \| a_i \|^2.
\end{aligned}
\tag{22}
$$

This minimization process simplifies the problem to a quadratic optimization, solved with LU decomposition. The piecewise nature is represented by the term $K_d \in \mathcal{N}(\mathbf{x}_q)$, defined as samples along γ. The first component of Eq.(22) is a penalty term to minimize the geodesic distance between samples $\mathbf{x}_{i,j}$ and the regressed curve, where w_i are weight variables based on sample distances. This helps regress a curve that will lie close to $\mathbf{x}_{i,j}$, shifted by $\mathbf{x}_q$ in order to have the initial reconstructions co-registered. The second term represents the velocity of the curve (defined by v_i, approximating $\dot{\gamma}(t_i)$), minimizing the L_2 distance of the 1st derivative of γ. By minimizing the value of the curve's first derivatives, this prohibits any discontinuities or rapid transitions of the curve's direction, and is modulated by α_i. Finally, an acceleration penalty term (defined by a_i) focuses on the 2nd derivative of γ with respect to t_i by minimizing the L_2 norm. The acceleration is modulated by β_i. Estimates for v_i and a_i (weighted by $\{\lambda, \mu\}$, respectively), are generated using geometric finite differences. These estimates dictates the forward and backward step-size on the regressed curve, leading to directional vectors in $\mathcal{M}$ as shown in [22]. In order to minimize $E(\gamma)$, a nonlinear conjugate gradient technique defined in the low-dimensional space $\mathbb{R}^d$ is used, thus avoiding convergence and speed issues. The regressed curve γ is therefore defined for all time points, originating at t_0. The curve creates a group average of spatiotemporal transformations based on individual correction trajectories.

4.3. Prediction of spine correction

Finally, to predict the evolution of spine correction from an unseen preoperative spine model, we use the geodesic curve $\gamma : \mathbb{R}^D \to \mathcal{M}$ modeling the spatiotemporal changes of the spine, where each point $\mathbf{x} \in \mathcal{M}$ is associated to a speed vector $\mathbf{v}$ defined with a tangent plane on the manifold such that $\mathbf{v} \in \mathrm{T}_{\mathbf{x}}\mathcal{M}$.

Based on Riemannian theory, an exponential mapping function at $\mathbf{x}$ with velocity $\mathbf{v}$ can be defined from the geodesics such that $\mathrm{e}_{\mathbf{x}}^{\mathcal{M}}(\mathbf{v})$. Using this concept, parallel transport curves defined in $\mathrm{T}_{\mathbf{x}}$ can help define a series of time-index vectors along γ as proposed by [10]. The collection of parallel transport curves allows to generate an average trajectory in ambient space $\mathbb{R}^D$, describing the spine changes due to the corrective forces of tethering. The general goal is to begin the process at the preoperative sample, and navigate the piecewise-geodesic curve describing correction evolution in time, where one can extract the appearance at any point (time) in $\mathbb{R}^D$ using the exponential mapping. For implementation purposes, the parallel transport curve are constrained within a smooth tubular boundary perpendicular to the curve (from an ICA) to generate the spatiotemporal evolution in the coordinate system of the preoperative model.

Hence, given the manifold at time t_0 with $\mathbf{v}$ defined in the tangent plane and the regressed piecewise-geodesic curve γ, the parallel curve is obtained as:

$$\eta^{\mathbf{v}}(\gamma, s) = \mathrm{e}_{\gamma(s)}^{\mathcal{M}}\left(\mathbf{x}_{\gamma, t_0, s}(\mathbf{v})\right), \; s \in \mathbb{R}^d. \tag{23}$$

Therefore by repeating this mapping for manifold points seen as samples of individual progression trajectories along $\gamma(s)$, an evolution model can be generated. Whenever a new sample is embedded, new samples points along $\gamma(s)$, denoted as $\eta^{\mathbf{v}}(\gamma, \cdot)$ can be generated parallel to the regressed piecewise curve in $\mathcal{M}$, capturing the spatiotemporal changes in correction.

A time warp function allowing s to vary along the geodesic curve is described as $\phi_i(t) = \theta_i(t - t_0 - \tau_i) + t_0$. Here, we propose to incorporate a personalized acceleration factor based on the spine maturity and flexibility derived from the spine bending radiographs and Risser grade. A coefficient $\theta_i = C_i \times R_i$ describing the change in Cobb angle C_i between poses, and modulated by the Risser grade R_i. This coefficient regulates the rate of correction based on the K neighboring samples. Finally, to take under account the relative differences between the group-wise samples and the query model once mapped onto the regressed curve, a time-shift parameter τ_i is incorporated in the warp function.

For spine correction evolution, displacement vectors $\mathbf{v}_i$ are obtained by a PCA of the hyperplane crossing $T_{\mathbf{x}_i}\mathcal{M}$ in manifold $\mathcal{M}$ [10]. Hence, for any query sample $\mathbf{x}_q$ which represents the mapped preoperative 3D reconstruction (prior to surgery), the predicted model at time t_k can be regressed from the piecewise-geodesic curve generated from embedded samples $\mathbf{x}$ in $\mathcal{N}(\mathbf{x}_q)$ such that:

$$\mathbf{y}_{q,t_k} = \eta^{\mathbf{v}_q}(\gamma, \phi_i(t_k)) + \varepsilon_{q,t_k} \tag{24}$$

which yields a predicted postoperative model $\mathbf{y}_{q,t_k}$ in high-dimensional space $\mathbb{R}^D$, and ε_{q,t_k} a zero-mean Gaussian distribution. The generated model offers a complete constellation of interconnected vertebral models composing the spine shape $\mathbf{S}$, at first-erect (FE), 1 or 2-year visits, including landmarks on vertebral endplates and pedicle extremities, which can be used to capture the local shape morphology with the correction process.

4.4. Experiments

The discriminant manifold was trained from a database of 438 3D spine reconstructions generated from biplanar images [23], originating from 131 patients demonstrating several types of deformities with immediate follow-up (FE), 1 and 2 year visits. Patients were recruited from a single center prospective study. Patients were divided in two groups, with the first group composed of 94 responsive patients showing a reduction in Cobb angle over or equal to 10° between the FE and follow-up visit. The second group was composed of 37 nonresponsive (NP) patients with a reduction of less than 10°. We evaluated the geometrical accuracy of the predictive manifold for 56 unseen surgical patients (mean age 12 ± 3, average main Cobb

	FE visit			**1-year visit**			**2-year visit**		
	3D RMS	**Dice**	**Cobb**	**3D RMS**	**Dice**	**Cobb**	**3D RMS**	**Dice**	**Cobb**
Biomec. sim	3.3 ± 1.1	85 ± 3.4	2.8 ± 0.8	3.6 ± 1.2	84 ± 3.6	3.2 ± 0.9	4.1 ± 2.3	82 ± 3.9	3.6 ± 1.0
LL-LVM [20]	3.6 ± 1.4	83 ± 4.0	3.8 ± 1.5	4.7 ± 3.3	79 ± 4.4	5.5 ± 2.6	6.6 ± 4.4	71 ± 5.9	7.0 ± 3.9
Deep AE [24]	4.1 ± 1.5	80 ± 4.4	5.1 ± 2.7	5.0 ± 1.9	77 ± 4.9	5.8 ± 3.0	6.3 ± 4.6	72 ± 5.7	6.6 ± 4.2
Proposed	2.4 ± 0.8	92 ± 2.7	1.8 ± 0.5	2.9 ± 0.9	90 ± 2.8	2.0 ± 0.7	3.2 ± 1.3	87 ± 3.1	2.1 ± 0.6

Predictions are evaluated at FE, 1 and 2-years.

Table 1. 3D RMS errors (mm), dice (%) and cobb angles (°) for the proposed method, and compared with biomechanical simulations, locally linear latent variable models (LL-LVM) and deep auto-encoders (AE).

angle on the frontal plane at the first visit was $47 \pm 10^{\circ}$), with predictions at $t = 0$ (FE), $t = 12$ and $t = 24$ months. For the predicted models, we evaluated the 3D root-mean-square difference of the vertebral landmarks generated, the Dice coefficients of the vertebral shapes and in the main Cobb angle. The results are shown in **Table 1**. Results were confronted to other techniques such as biomechanical simulations performed on each subject using finite element modeling with ex-vivo parameters [25], a locally linear latent variable model [20] and a deep auto-encoder network [24]. Results from the predicted geometrical models show the regressed spatiotemporal geodesic curve yields anatomically coherent structures, with accurate local vertebral morphology.

5. Discussion

Algorithms capable of extracting clinically relevant and meaningful descriptions from medical imaging datasets have become of widespread interest to theoreticians as well as practitioners in the medical field, accelerating the pace in recent years involving varied fields such as in machine learning, geometry, statistics and genomics to propose new insights for the analysis of imaging and biologic datasets. Towards this end, manifold learning has demonstrated a tremendous potential to learn the underlying representation of high-dimensional, complex imaging datasets.

We presented frameworks describing longitudinal, multimodal image features from neuroimaging data using a Bayesian model for discriminant nonlinear manifolds to predict the conversion of progressive MCI to Alzheimer's disease. This probabilistic method introduces class-dependent latent variables which is based on the concept that local structure is transformed from manifold to the high-dimensional domain. This variational learning method can ultimately assess uncertainty within the manifold domain, which can lead to a better understanding of relationships between converters and nonconverters for patients with MCI.

Finally, a prediction method for the outcomes of spine surgery using geodesic parallel transport curves generated from probabilistic manifold models was presented. The mathematical models allow to describe patterns in a nonlinear and discriminant Riemannian framework by first distinguishing nonprogressive and progressive cases, followed by a prediction of structural evolution. The proposed model provides a way to analyze longitudinal samples from a geodesic curve in manifold space, thus simplifying the mixed effects when studying group-average trajectories.

Author details

Samuel Kadoury

Address all correspondence to: samuel.kadoury@polymtl.ca

Polytechnique Montreal, Succ. Centre-Ville, Montreal, QC, Canada

References

[1] Yang Y, Dunson DB, et al. Bayesian manifold regression. The Annals of Statistics. 2016; **44**(2):876-905

[2] Davatzikos C, Resnick SM, Wu X, Parmpi P, Clark CM. Individual patient diagnosis of AD and FTD via high-dimensional pattern classification of MRI. NeuroImage. 2008;**41**(4): 1220-1227

[3] Li S, Shi F, Pu F, Li X, Jiang T, Xie S, Wang Y. Hippocampal shape analysis of Alzheimer disease based on machine learning methods. American Journal of Neuroradiology. 2007; **28**(7):1339-1345

[4] Beg MF, Miller MI, Trouvé A, Younes L. Computing large deformation metric mappings via geodesic flows of diffeomorphisms. International Journal of Computer Vision. 2005; **61**(2):139-157

[5] Fletcher PT, Lu C, Joshi S. Statistics of shape via principal geodesic analysis on lie groups. In: Computer Vision and Pattern Recognition, 2003. Proceedings. 2003 IEEE Computer Society Conference on. Vol. 1. IEEE; 2003. pp. I-I

[6] Pennec X. Intrinsic statistics on Riemannian manifolds: Basic tools for geometric measurements. Journal of Mathematical Imaging and Vision. 2006;**25**(1):127

[7] Fletcher PT, Venkatasubramanian S, Joshi S. The geometric median on Riemannian manifolds with application to robust atlas estimation. NeuroImage. 2009;**45**(1):S143-S152

[8] Singh N, Hinkle J, Joshi S, Fletcher PT. A hierarchical geodesic model for diffeomorphic longitudinal shape analysis. In: International Conference on Information Processing in Medical Imaging. Springer; 2013. pp. 560-571

[9] Fishbaugh J, Prastawa M, Gerig G, Durrleman S. Geodesic regression of image and shape data for improved modeling of 4D trajectories. In: 2014 International Symposium on Biomedical Imaging. IEEE; 2014. pp. 385-388

[10] Schiratti JB, Allassonniere S, Colliot O, Durrleman S. Learning spatiotemporal trajectories from manifold-valued longitudinal data. In: Advances in Neural Information Processing Systems. 2015:2404-2412

[11] Kadoury S, Mandel W, Roy-Beaudry, Nault ML, Parent S. 3-D morphology prediction of progressive spinal deformities from probabilistic modeling of discriminant manifolds. IEEE Transactions on Medical Imaging. 2017;**36**(5):1194-1204

[12] Chevallier J, Oudard S, Allassonnière S. Learning spatiotemporal piecewise-geodesic trajectories from longitudinal manifold-valued data. In: 31st Conference on Neural Information Processing Systems (NIPS 2017), Long Beach, CA, USA; 2017

[13] Rekik I, Li G, Lin W, Shen D. Predicting infant cortical surface development using a 4D varifold-based learning framework and local topography-based shape morphing. Medical Image Analysis. 2016;**28**:1-12

[14] Kadoury S, Labelle H, Paragios N. Spine segmentation in medical images using manifold embeddings and higher-order MRFs. IEEE Transactions on Medical Imaging. 2013;**32**: 1227-1238

[15] Roweis ST, Saul LK. Nonlinear dimensionality reduction by locally linear embedding. Science. 2000;**290**:2323-2326

[16] Nadaraya EA. On estimating regression. Theory of Probability and its Applications. 1964; **10**:186-190

[17] Davis B, Fletcher P, Bullitt E, Joshi S. Population shape regression from random design data. In: Proceedings of the 2007 IEEE Computer Society Conference on Computer Vision and Pattern Recognition, 2007. IEEE. 2007;**1**:1-8

[18] Rother C, Kohli P, Feng W, Jia J. Minimizing sparse higher order energy functions of discrete variables. In: Conference on Computer Vision and Pattern Recognition; 2009. pp. 1382-1389

[19] Komodakis N, Tziritas G, Paragios N. Performance vs computational efficiency for optimizing single and dynamic MRFs: Setting the state of the art with primal dual strategies. Computer Vision and Image Understanding. 2008;**112**(1):14-29

[20] Park M, Jitkrittum W, Qamar A, Szabó Z, Buesing L, Sahani M. Bayesian manifold learning: The locally linear latent variable model (LL-LVM). In: Advances in Neural Information Processing Systems. 2015:154-162

[21] Patenaude B, Smith SM, Kennedy DN, Jenkinson M. A Bayesian model of shape and appearance for subcortical brain segmentation. NeuroImage. 2011;**56**(3):907-922

[22] Boumal N, Absil PA. A discrete regression method on manifolds and its application to data on SO (n). IFAC Proceedings Volumes. 2011;**44**(1):2284-2289

[23] Humbert L, de Guise J, Aubert B, Godbout B, Skalli W. 3D reconstruction of the spine from biplanar X-rays using parametric models based on transversal and longitudinal inferences. Medical Engineering & Physics. 2009;**31**(6):681-687

[24] Thong W, Parent S, Wu J, Aubin CE, Labelle H, Kadoury S. Three-dimensional morphology study of surgical adolescent idiopathic scoliosis patient from encoded geometric models. European Spine Journal. 2016;**25**(10):3104-3113

[25] Cobetto N, Parent S, Aubin CE. 3D correction over 2 years with anterior vertebral body growth modulation: A finite element analysis of screw positioning, cable tensioning and postop functional activities. Clinical Biomechanics. 2018;**51**:26-33

Chapter 5

Trajectory Tracking Control of Parallel Manipulator with Integral Manifold and Observer

Zhengsheng Chen

Additional information is available at the end of the chapter

http://dx.doi.org/10.5772/intechopen.80375

Abstract

In view of the problem that notable flexible displacement will occur for parallel manipulators when operating at high speed, the composite controller based on the integral manifold and high-gain observer is proposed for trajectory tracking and the 3RRR parallel manipulator is taken as the object. Based on the stiffness matrix, the small variable is introduced to decompose the rigid-flexible coupling dynamic model into slow and fast subsystem. For the slow subsystem, the backstepping control is applied for rigid motion tracking. In order to account for the links' flexible displacement the corrective torque is deduced, and the compensation for the flexible displacement is realized. For the fast subsystem, the sliding mode control is utilized to suppress the vibration. The high-gain observer is designed to avoid the measurement of the curvature rate of flexible links. Also, the stability of the overall system is proven with the Lyapunov stability theorem and the upper bound of the small variable is obtained. At last, the proposed composite controller together with the singular perturbation control and the rigid body model-based backstepping control are simulated, and vibration suppression and tracking performances are compared to validate the proposed control scheme.

Keywords: parallel manipulator, integral manifold, high-gain observer, composite control, sliding mode control, backstepping control, vibration suppression

1. Introduction

Parallel manipulators (PMs) possess advantages of high precision, high stiffness, and large load-to-weight ratio; they have attracted wide attention and have been widely used in industries such as high-speed handling, motion simulation, and electronic manufacturing [1]. However, in order to increase efficiency, PMs are increasingly used in high-speed and heavy-duty operations. In order to reduce costs and energy consumption, the lightweight design of the mechanical body will be the inevitable choice. However, in the high-speed or heavy-duty application, the lightweight mechanical body will produce significant elastic deformation and vibration. Therefore, the end-effector's movement consists of the rigid-body motion and the elastic displacement caused by elastic deformation and vibration. Using conventional control methods for rigid-body manipulators will not guarantee good tracking accuracy of flexible manipulator's end-effector. Therefore, it is of great significance to improve the tracking accuracy of high-speed lightweight PMs by considering the flexibility of members to establish the dynamic model for rigid-flexible coupling and carrying out research on high-precision control algorithms.

Many scholars have conducted extensive and in-depth studies on modeling methods for manipulators with flexible links. Dwivedy et al. [2] reviewed the dynamic modeling of robots with flexible links. Due to the presence of link flexibility, the system will exhibit nonminimum phase characteristics when selecting the end-effector of the manipulator as the output. The literature [3–5] redefines the output of the manipulator's end position by taking the link elasticity into account, and uses the control algorithm for the rigid-body manipulator to control the new output; however, this method can only realize the point-to-point position control and cannot guarantee tracking control of the end trajectory [6]. The singular perturbation method is another effective method to deal with the nonminimum phase characteristics of manipulators with elastic links. The small parameters are introduced to reduce the order of rigid-flexible coupling models, which are decomposed into two subsystems, the fast and the slow, and two subcontrollers are designed using compound control algorithm. The controller of the system realizes the control of the rigid body motion and the rapid suppression of the elastic vibration. However, as the deformation increases, the singularity perturbation algorithm shows a deficiency and the algorithm cannot compensate for the elastic displacement [7–9]. Khorasani [10] proposed an integral manifold method by high-order approximation of fast subsystem variables, which greatly improved the vibration suppression effect. By introducing the elastic displacement into the end of the manipulator and designing the corrective torque, Moallem et al. [11] realized the trajectory tracking precision control and vibration suppression of the two-degree-of-freedom serial robot. Based on the above method, Fotouhi et al. [12–16] studied the trajectory tracking control of the flexible joint robot, the flexible robot with the single link, the rigid-flexible hybrid robot, and the two-bar flexible robot by simplifying the selection of correction moments, and show good results.

Due to the existence of the closed-chain structure, the dynamic model of PMs is complex when considering the flexibility of the links. Therefore, the research on the vibration suppression and trajectory tracking control is very limited. Zhang et al. [7] used assumption mode method and Lagrange equation to model 3PRR PMs with flexible passive links, and adopted singular perturbation compound control to suppress vibration. However, the influence of the elastic displacement of the links on the moving platform is not considered in the model, and the elastic displacement compensation and the rate of change of the elastic links are not processed when the algorithm is designed. Therefore, the trajectory tracking effect needs to be improved. Existing research has not yet been found for the above issues. In the research of trajectory tracking control based on integral manifold, no relevant research has been found for PMs. The control algorithms for the slow subsystem in the existing research are feedback linearization methods, and the fast subsystem is PD control or pole placement. In order to taking into accounts of the elastic deformation and vibration of high-speed PMs due to the flexibility of links and improve the tracking accuracy and dynamic performance, this chapter introduces the integral manifold based on the rigid-flexible coupling model of the 3RRR PM, the hypothesis of small deformation and the velocity mapping in the previous paper [17], and the high-order rigid-flexible coupling model is transformed into two subsystems, then a composite control algorithm based on sliding mode variable structure control and backstepping control is proposed. At the same time, a high-gain observer is introduced to the curvature rate caused by the flexibility. Finally, simulation studies are conducted to verify the feasibility of the algorithm.

2. The dynamic model of the 3RRR PM

The structure of the 3RRR parallel manipulator was shown in **Figure 1**, which consists of three branches, and each branch composed of one active link and passive link, the end of which is the moving platform. The coordinates and the parameters are given in **Figure 2**, $O-XY$ and $G-x_G y_G$ are the coordinate frames attached with the base and moving platform, with O and G as the origin, respectively. θ_i and β_i are the angles of the active and passive links, $i=1,2,3$, the position and attitude of the moving platform are depicted as $\boldsymbol{\eta}=\begin{bmatrix}x & y & \phi\end{bmatrix}^{\mathrm{T}}$ in the base frame.

According to our previously published paper [17], the flexibility of passive links can be neglected, so only the deformation of active links is considered here, which can be expressed as $\delta_i=\sum_{k=1}^{n}\alpha_i^k m_i^k, i=1,2,3$, where α_i^k and m_i^k are the shape function and the curvature of the kth point in the ith active link, respectively, where $k=1$. According to [17], after ignoring the deformation of the passive links and adding the parameters of the motors and reducers, the dynamic model of the PM can be expressed as:

$$\begin{bmatrix}(\boldsymbol{M}_{11})_0+(\boldsymbol{M}_{11})_1 & \boldsymbol{M}_{12}\\ \boldsymbol{M}_{12}^{\mathrm{T}} & \boldsymbol{M}_{22}\end{bmatrix}\begin{bmatrix}\ddot{\boldsymbol{\eta}}\\ \ddot{\boldsymbol{m}}\end{bmatrix}+\begin{bmatrix}\boldsymbol{0} & \boldsymbol{0}\\ \boldsymbol{0} & \boldsymbol{K}\end{bmatrix}\begin{bmatrix}\boldsymbol{\eta}\\ \boldsymbol{m}\end{bmatrix}+\begin{bmatrix}(\boldsymbol{f}_1)_0+\boldsymbol{M}_{f1}\boldsymbol{m}+\boldsymbol{M}_{f2}\dot{\boldsymbol{m}}\\ (\boldsymbol{f}_2)_0+\boldsymbol{M}_{f3}\dot{\boldsymbol{m}}\end{bmatrix}=\begin{bmatrix}\boldsymbol{J}_{p\theta}^{\mathrm{T}}\boldsymbol{\tau}\\ \boldsymbol{0}\end{bmatrix} \quad (1)$$

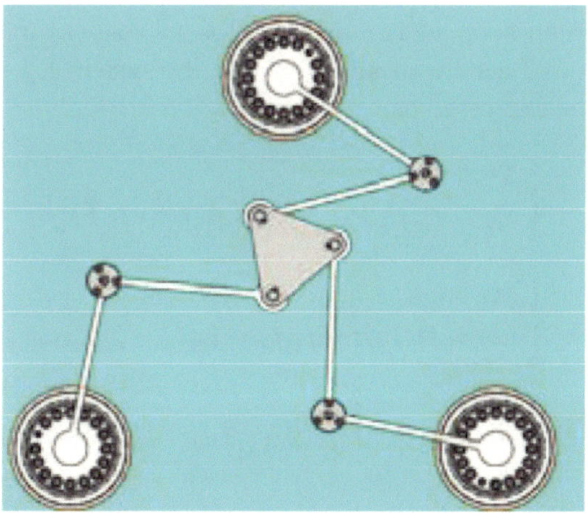

Figure 1. The 3RRR PM.

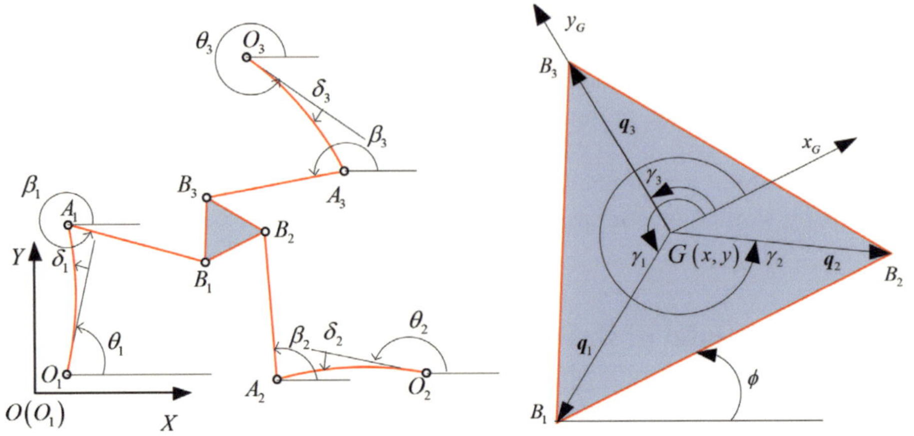

Figure 2. Coordinates of the 3RRR PM.

where J_m and J_g are the moment of inertia of the motor and the reducer, and $\boldsymbol{K} = \mathbf{diag}([k_s, k_s, k_s])$ is the stiffness matrix, while k_s and i_g are the link's stiffness and the reduction ratio, respectively, $\boldsymbol{\tau}$ represents the driving torque, $(\boldsymbol{M}'_{11})_0$ and $(\boldsymbol{f}'_1)_0$ are the mass matrix and quadratic terms in the dynamic equation derived from [17], while the item corresponding to $\boldsymbol{m}$ is neglected.

3. Integral manifold-based model reduction of the high-speed PM

From the dynamic model (1), the state variables are defined as below [15],

$$\begin{cases} \boldsymbol{X}_1 = \boldsymbol{\eta}, \boldsymbol{X}_2 = \dot{\boldsymbol{\eta}} \\ z_1 = \boldsymbol{m}/\varepsilon^2, z_2 = \dot{\boldsymbol{m}}/\varepsilon \end{cases} \tag{2}$$

where $\boldsymbol{z} = [z_1\ z_2]^{\mathrm{T}}$ and $\boldsymbol{X} = [\boldsymbol{X}_1\ \boldsymbol{X}_2]^{\mathrm{T}}$ are state variables of the slow subsystem, $\varepsilon \in \boldsymbol{R}$ is the small parameter larger than zero, which are used for subsequent model reduction and time scale transformation. From the state variables (2) and the system Eq. (1), the state equation of the perturbed form can be expressed as:

$$\begin{cases} \dot{\boldsymbol{X}}_1 = \boldsymbol{X}_2, \\ \dot{\boldsymbol{X}}_2 = \boldsymbol{J}_{11}\boldsymbol{J}_{p\theta}^{\mathrm{T}}\boldsymbol{\tau} - \boldsymbol{J}_{11}\boldsymbol{f}_1 - \boldsymbol{J}_{12}\boldsymbol{f}_2 - \boldsymbol{J}_{12}\tilde{k}z_1; \end{cases} \tag{3}$$

$$\begin{cases} \varepsilon\dot{z}_1 = z_2, \\ \varepsilon\dot{z}_2 = \boldsymbol{J}_{12}^{\mathrm{T}}\boldsymbol{J}_{p\theta}^{\mathrm{T}}\boldsymbol{\tau} - \boldsymbol{J}_{12}^{\mathrm{T}}\boldsymbol{f}_1 - \boldsymbol{J}_{22}\boldsymbol{f}_2 - \boldsymbol{J}_{22}\tilde{k}z_1. \end{cases} \tag{4}$$

where $\tilde{k} = k_s\varepsilon^2$ is the stiffness coefficient, $\boldsymbol{J} = \begin{bmatrix} \boldsymbol{J}_{11} & \boldsymbol{J}_{12}; & \boldsymbol{J}_{12}^{\mathrm{T}} & \boldsymbol{J}_{22} \end{bmatrix}$ is the inverse matrix of the mass matrix $\boldsymbol{M}$.

For Eq. (4), the integral manifold is defined as [15, 18],

http://dx.doi.org/10.5772/intechopen.80375

$$z(t^*,\varepsilon) = \boldsymbol{h}(\boldsymbol{X}_1(t^*,\varepsilon), \boldsymbol{X}_2(t^*,\varepsilon), \boldsymbol{\tau}(t^*),\varepsilon) \Rightarrow z(t,\varepsilon) = \boldsymbol{h}^a(\boldsymbol{X}_1(t,\varepsilon), \boldsymbol{X}_2(t,\varepsilon), \boldsymbol{\tau}(t),\varepsilon) \tag{5}$$

Eq. (5) can be interpreted that if the fast subsystem variables arrive at the integral manifold trajectory at the moment t^*, then for the moment $\forall t > t^*$, the variable will always remain on the manifold trajectory. In order to ensure the above conditions valid, the additional control variables are added in the control system.

Due to the small variable ε close to 0, the integral manifold $\boldsymbol{h}$ and the moment $\boldsymbol{\tau}$ are all functions of ε, Taylor expansion of the above variables is available as,

$$\begin{cases} \boldsymbol{h}_1^a \approx \boldsymbol{h}_1 = \boldsymbol{h}_{10} + \varepsilon \boldsymbol{h}_{11}(\boldsymbol{X}_1, \boldsymbol{X}_2, t) + \ldots + \varepsilon^p \boldsymbol{h}_{1p}(\boldsymbol{X}_1, \boldsymbol{X}_2, t) \\ \boldsymbol{h}_2^a \approx \boldsymbol{h}_2 = \boldsymbol{h}_{20} + \varepsilon \boldsymbol{h}_{21}(\boldsymbol{X}_1, \boldsymbol{X}_2, t) + \ldots + \varepsilon^p \boldsymbol{h}_{2p}(\boldsymbol{X}_1, \boldsymbol{X}_2, t) \\ \boldsymbol{\tau} \approx \boldsymbol{\tau}_0 + \varepsilon \boldsymbol{\tau}_1(\boldsymbol{X}_1, \boldsymbol{X}_2, t) + \ldots + \varepsilon^p \boldsymbol{\tau}_p(\boldsymbol{X}_1, \boldsymbol{X}_2, t). \end{cases} \tag{6}$$

where $\boldsymbol{h}_1$ and $\boldsymbol{h}_2$ are the approximations of ${\boldsymbol{h}_1}^a$and ${\boldsymbol{h}_2}^a$, and $\boldsymbol{h}_{ij} = \frac{\partial^j \boldsymbol{h}_i^a}{j!\partial\varepsilon^j}|_{\varepsilon=0}$ is the derivative of the integral manifold with respect to the small variable ε, while $i = 1, 2$ $j = 0, 1, 2, \ldots p$, and $p \in N^+$ is the approximation order. Since the elastic displacement of the link is ε^2 times of the state variable z of the fast subsystem, so p should be at least 2 when the elastic displacement can be accounted in the end trajectory, the p is selected 2 here.

The inverse matrix of the mass matrix, the Coriolis force and the centrifugal force terms are functions of the small variable ε, the Taylor expansion of the inverse matrix about ε can be expressed as,

$$\begin{cases} J_{11} = (J_{11})_0, J_{12} = (J_{12})_0 \\ J_{22} = (J_{22})_0 + (J_{22})_2 \varepsilon^2/2 \end{cases} \tag{7}$$

The centrifugal and inertial force after the expansion of Eq. (1) can be expressed as,

$$\begin{cases} f_1 = (f_1)_0 + \left((f_1)_{20}\boldsymbol{h}_{10} + (f_1)_{21}\dot{\boldsymbol{h}}_{10}\right)\varepsilon^2/2, \\ f_2 = (f_2)_0 + \varepsilon^2 (f_2)_{21}\dot{\boldsymbol{h}}_{10}/2. \end{cases} \tag{8}$$

Substituting Eqs. (6) through (8) into Eq. (4), we can obtain,

$$\begin{cases} \boldsymbol{h}_{10} = \left(J_{22}\tilde{k}\right)_0^{-1} \left(J_{12}^{\mathrm{T}} J_{p\theta}^{\mathrm{T}} \boldsymbol{\tau}_0 - J_{12}^{\mathrm{T}} (f_1)_0 - (J_{22})_0 (f_2)_0\right), \\ \boldsymbol{h}_{11} = \left(J_{22}\tilde{k}\right)_0^{-1} \left(J_{12}^{\mathrm{T}} J_{p\theta}^{\mathrm{T}} \boldsymbol{\tau}_1 - \dot{\boldsymbol{h}}_{20}\right), \\ \boldsymbol{h}_{12} = \left(J_{22}\tilde{k}\right)_0^{-1} \left(J_{12}^{\mathrm{T}} J_{p\theta}^{\mathrm{T}} \boldsymbol{\tau}_2 - \dot{\boldsymbol{h}}_{21} - J_{12}^{\mathrm{T}}\left((f_1)_{20}\boldsymbol{h}_{10} + (f_1)_{21}\dot{\boldsymbol{h}}_{10}\right)/2 - \right)(J_{22})_2\left((f_2)_0 + \boldsymbol{h}_{10}\right)/2 - (J_{22})_0 (f_2)_{21}\dot{\boldsymbol{h}}_{10}/2\Big), \\ \boldsymbol{h}_{20} = 0, \boldsymbol{h}_{21} = \dot{\boldsymbol{h}}_{10}, \boldsymbol{h}_{22} = \dot{\boldsymbol{h}}_{11}. \end{cases} \tag{9}$$

When the flexibility of the links is ignored, the small variable $\varepsilon = 0$ is valid. Substituting $\boldsymbol{h}_1$ into Eq. (3), the differential equation of the slow subsystem can be obtained as,

$$\begin{cases} \dot{\overline{X}}_1 = \overline{X}_2 \\ \dot{\overline{X}}_2 = (M_{11})_0^{-1} J_{p\theta}^{\mathrm{T}} \tau_0 - (M)_{110}^{-1} (f_1)_0 \end{cases} \tag{10}$$

where $\overline{X}_1$ and $\overline{X}_2$ represent variables of the slow subsystem, for the convenience of description, $\overline{X}_1$ and $\overline{X}_2$ are replaced by X_1 and X_2 in the following expressions,

According to the integral manifold, the deviation of the fast subsystem variable can be expressed as,

$$\begin{cases} X_{f1} = z_1 - h_{10} - \varepsilon h_{11} - \varepsilon^2 h_{12} \\ X_{f2} = z_2 - h_{20} - \varepsilon h_{21} - \varepsilon^2 h_{22} \end{cases} \tag{11}$$

Multiply the Eq. (11) with ε, derive and substitute it into Eq. (6). According to Eq. (9), the fast subsystem equation can be obtained by substituting h_{ij},

$$\begin{cases} \varepsilon \dot{X}_{f1} = X_{f2}, \\ \varepsilon \dot{X}_{f2} = J_{12}^{\mathrm{T}} J_{p\theta}^{\mathrm{T}} \tau_f - (J_{22})_0 \tilde{k} X_{f1} - \varepsilon^2 \left((J_{22})_2 + J_{12}^{\mathrm{T}} (f_1)_{20} \right) X_{f1}/2 - \\ \qquad \varepsilon \left(J_{12}^{\mathrm{T}} (f_1)_{21} + (J_{22})_0 (f_2)_{21} \right) X_{f2}/2. \end{cases} \tag{12}$$

For the slow and fast subsystems represented by Eqs. (10) and (12), the composite control algorithm is designed as shown in **Figure 3**. For the slow subsystem, the backstepping control is used to achieve the tracking control of the rigid body motion. At the same time, according to the velocity mapping relationship, the mapping relationship between the elastic deformation of the links and the elastic displacement of the moving platform is established. The motion of the moving platform is obtained according to the rigid-body motion and the elastic displacement, and the elastic torque compensation is realized by designing the correction torque τ_1 and τ_2. For fast subsystems, the sliding mode control is used to ensure the manifold valid. Considering the difficulty of measuring the rate of curvature change of the links, a high-gain observer will be designed to estimate the rate of curvature change based on the curvature value. The algorithm design will be based on the control structure shown in **Figure 3**.

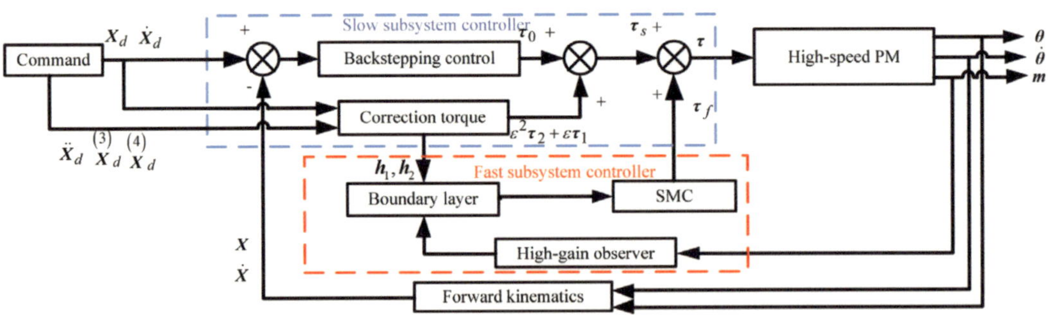

Figure 3. Scheme of the controller.

4. The backstepping algorithm-based slow subsystem control

The backstepping control is a recursive control algorithm for complex nonlinear systems. The original system is decomposed into subsystems that do not exceed the system order. The control design is realized by establishing Lyapunov functions step by step for each subsystem, and the stability of the system is ensured [19]. First, define the position error as,

$$\boldsymbol{e}_1 = \boldsymbol{X}_1 - \boldsymbol{X}_d \tag{13}$$

where $\boldsymbol{X}_d$ is the command signal, define the amount of virtual control as,

$$\upsilon_1 = -c_1\boldsymbol{e}_1 + \dot{\boldsymbol{X}}_d \tag{14}$$

where c_1 is a constant greater than zero and the velocity error $\boldsymbol{e}_2$ can be defined as,

$$\boldsymbol{e}_2 = \boldsymbol{X}_2 - \upsilon_1 . \tag{15}$$

Based on the position error, define the Lyapunov function as,

$$V_1 = 1/2\boldsymbol{e}_1^{\mathrm{T}}\boldsymbol{e}_1 . \tag{16}$$

Deriving Eq. (16) can be obtained,

$$\dot{V}_1 = \boldsymbol{e}_1^{\mathrm{T}}\dot{\boldsymbol{e}}_1 = -c_1\boldsymbol{e}_1^{\mathrm{T}}\boldsymbol{e}_1 + \boldsymbol{e}_1^{\mathrm{T}}\boldsymbol{e}_2 . \tag{17}$$

According to the velocity error (15) in conjunction with Eq. (17), the Lyapunov function is defined as,

$$V_2 = V_1 + 1/2\boldsymbol{e}_2^{\mathrm{T}}\boldsymbol{e}_2 = 1/2\boldsymbol{e}_1^{\mathrm{T}}\boldsymbol{e}_1 + 1/2\boldsymbol{e}_2^{\mathrm{T}}\boldsymbol{e}_2 . \tag{18}$$

Deriving the above formula and substituting the relevant parameters, the derivative of the Lyapunov function can be expressed as,

$$\dot{V}_2 = -c_1\boldsymbol{e}_1^{\mathrm{T}}\boldsymbol{e}_1 + \boldsymbol{e}_1^{\mathrm{T}}\boldsymbol{e}_2 + \boldsymbol{e}_2^{\mathrm{T}}\dot{\boldsymbol{e}}_2 = \boldsymbol{e}_2^{T}\left((\boldsymbol{M}_{11})_0^{-1}\boldsymbol{J}_{p\theta}^{\mathrm{T}}\boldsymbol{\tau}_0 - (\boldsymbol{M}_{11})_0^{-1}(\boldsymbol{f}_1)_0 + c_1\dot{\boldsymbol{e}}_1 - \ddot{\boldsymbol{X}}_d\right) - c_1\boldsymbol{e}_1^{\mathrm{T}}\boldsymbol{e}_1 + \boldsymbol{e}_1^{\mathrm{T}}\boldsymbol{e}_2 \tag{19}$$

According to Eq. (19), the control torque of the slow subsystem is

$$\begin{aligned}\boldsymbol{\tau}_0 &= \left(\boldsymbol{J}_{p\theta}^{\mathrm{T}}\right)^{-1}\left((\boldsymbol{f}_1)_0 + (\boldsymbol{M}_{11})_0(-c_1\dot{\boldsymbol{e}}_1 + \ddot{\boldsymbol{X}}_d - c_2\boldsymbol{e}_2 - \boldsymbol{e}_1)\right) \\ &= \left(\boldsymbol{J}_{p\theta}^{\mathrm{T}}\right)^{-1}\left((\boldsymbol{f}_1)_0 + (\boldsymbol{M}_{11})_0(\ddot{\boldsymbol{X}}_d - (c_1 + c_2)\dot{\boldsymbol{e}}_1 - (c_1c_2 + 1)\boldsymbol{e}_1)\right)\end{aligned} \tag{20}$$

where c_2 is a positive real number, and substitute Eq. (20) into Eq. (19), the derivative of the Lyapunov function of the slow subsystem can be expressed as:

$$\dot{V} = -c_1\boldsymbol{e}_1^{\mathrm{T}}\boldsymbol{e}_1 + c_2\boldsymbol{e}_2^{\mathrm{T}}\boldsymbol{e}_2 \leq 0 \tag{21}$$

Therefore, according to the Lyapunov stability principle, the slow subsystem is stable with the torque $\boldsymbol{\tau}_0$. Due to the existence of the elasticity, the end position of the PM can be expressed as:

$$\boldsymbol{r} = \boldsymbol{X}_1 + \boldsymbol{f}_3(\boldsymbol{\eta}, \boldsymbol{h}_{10}, \boldsymbol{h}_{11}, \boldsymbol{h}_{12}, \varepsilon). \tag{22}$$

where $\boldsymbol{f}_3$ is the elastic displacement of the center G of the moving platform induced by the elastic deformation and vibration of the links, which is the elastic displacement of the end-effector of the moving platform.

According to the velocity mapping relationship, the acceleration of the moving platform generated by the elastic motion can be expressed as,

$$\ddot{\boldsymbol{f}}_3 = \varepsilon^2 \boldsymbol{J}_{p\theta}^{-1} \phi_l (\boldsymbol{h}_{10} + \varepsilon \boldsymbol{h}_{11} + \varepsilon^2 \boldsymbol{h}_{12})/l_1 + \varepsilon^2 \dot{\overline{\boldsymbol{J}_{p\theta}^{-1}}} \phi_l \left(\dot{\boldsymbol{h}}_{10} + \varepsilon \dot{\boldsymbol{h}}_{11} + \varepsilon^2 \dot{\boldsymbol{h}}_{12} \right)/l_1. \tag{23}$$

where $\dot{\overline{\boldsymbol{J}_{p\theta}^{-1}}}$ is the time derivative of $\boldsymbol{J}_{p\theta}^{-1}$.

The flexibility examined in this chapter is within a small deformation range, and the elastic displacement $\boldsymbol{f}_3$ of the end-effector of the moving platform due to the elastic displacement of the rod can be simplified as,

$$\boldsymbol{f}_3 = \varepsilon^2 \boldsymbol{J}_{p\theta}^{-1} \phi_l \left(\boldsymbol{h}_{10} + \varepsilon \boldsymbol{h}_{11} + \varepsilon^2 \boldsymbol{h}_{12} \right)/l_1. \tag{24}$$

Make the second derivative of Eq. (22), when considering the rigid-flexible coupling motion, the acceleration of the end-effector of the moving platform can be expressed as,

$$\begin{aligned} \ddot{\boldsymbol{r}} &= \ddot{\boldsymbol{X}}_d + (c_1 + c_2)(\dot{\boldsymbol{X}}_d - \dot{\boldsymbol{\eta}}) + (c_1 c_2 + 1)(\boldsymbol{X}_d - \boldsymbol{\eta}) + \boldsymbol{M}_{11}^{-1} \boldsymbol{J}_{p\theta}^{\mathrm{T}} (\varepsilon \boldsymbol{\tau}_1 + \varepsilon^2 \boldsymbol{\tau}_2) \\ &+ \varepsilon^2 \left(\boldsymbol{J}_{p\theta}^{-1} \phi_l \ddot{\boldsymbol{h}}_{10} + \dot{\overline{\boldsymbol{J}_{p\theta}^{-1}}} \phi_l \dot{\boldsymbol{h}}_{10} \right)/l_1 - \boldsymbol{M}_{11}^{-1} \left((\boldsymbol{f}_1)_{20} \boldsymbol{h}_{10} + (\boldsymbol{f}_1)_{21} \dot{\boldsymbol{h}}_{10} \right) \varepsilon^2/2 \\ &+ \boldsymbol{J}_{12} (\boldsymbol{J}_{22})_0^{-1} (\boldsymbol{J}_{22})_2 \left((\boldsymbol{f}_2)_0 + \boldsymbol{h}_{10} \right) \varepsilon^2/2 + \boldsymbol{J}_{12} (\boldsymbol{J}_{22})_0^{-1} \ddot{\boldsymbol{h}}_{10} \varepsilon^2 + \boldsymbol{J}_{12} (\boldsymbol{f}_2)_{21} \dot{\boldsymbol{h}}_{10} \varepsilon^2/2. \end{aligned} \tag{25}$$

Defining the position error $\boldsymbol{e}_3 = \boldsymbol{X}_d - \boldsymbol{r}$ and velocity error $\boldsymbol{e}_4 = \dot{\boldsymbol{e}}_3$ of the end-effector of the moving platform, Eq. (25) can be transformed as,

$$\begin{cases} \dot{\boldsymbol{e}}_3 = \boldsymbol{e}_4 \\ \dot{\boldsymbol{e}}_4 = -(c_1 + c_2)\boldsymbol{e}_4 - (c_1 c_2 + 1)\boldsymbol{e}_3 - \boldsymbol{M}_{11}^{-1} \boldsymbol{J}_{p\theta}^{\mathrm{T}} (\varepsilon \boldsymbol{\tau}_1 + \varepsilon^2 \boldsymbol{\tau}_2) - \varepsilon^2 \boldsymbol{J}_{12} (\boldsymbol{J}_{22})_0^{-1} \ddot{\boldsymbol{h}}_{10} \\ \quad - \varepsilon^2 \left(\boldsymbol{J}_{p\theta}^{-1} \phi_l \ddot{\boldsymbol{h}}_{10} + \dot{\overline{\boldsymbol{J}_{p\theta}^{-1}}} \phi_l \dot{\boldsymbol{h}}_{10} \right)/l_1 + \varepsilon^2 \boldsymbol{M}_{11}^{-1} \left((\boldsymbol{f}_1)_{20} \boldsymbol{h}_{10} + (\boldsymbol{f}_1)_{21} \dot{\boldsymbol{h}}_{10} \right)/2 - \varepsilon^2 (c_1 c_2 + 1) \boldsymbol{J}_{p\theta}^{-1} \phi_l \boldsymbol{h}_{10}/l_1 \\ \quad - \varepsilon^2 \boldsymbol{J}_{12} (\boldsymbol{J}_{22})_0^{-1} (\boldsymbol{J}_{22})_2 \left((\boldsymbol{f}_2)_0 + \boldsymbol{h}_{10} \right)/2 - \varepsilon^2 (c_1 + c_2) \boldsymbol{J}_{p\theta}^{-1} \phi_l \dot{\boldsymbol{h}}_{10}/l_1 - \varepsilon^2 \boldsymbol{J}_{12} (\boldsymbol{f}_2)_{21} \dot{\boldsymbol{h}}_{10}/2 \end{cases} \tag{26}$$

According to Eq. (26), define the Lyapunov function,

$$\boldsymbol{V}_3 = 1/2\boldsymbol{e}_3^{\mathrm{T}}\left(c_1c_2+1\right)\boldsymbol{e}_3 + 1/2\boldsymbol{e}_4^{\mathrm{T}}\boldsymbol{e}_4. \tag{27}$$

Derivative of Eq. (27) with respect to time can be obtained as,

$$\begin{aligned}
\dot{\boldsymbol{V}}_3 &= 1/2\boldsymbol{e}_3^{\mathrm{T}}\left(c_1c_2+1\right)\dot{\boldsymbol{e}}_3 + 1/2\boldsymbol{e}_4^{T}\dot{\boldsymbol{e}}_4 = \\
&-1/2\boldsymbol{e}_4^{\mathrm{T}}\left(c_1+c_2\right)\boldsymbol{e}_4 + 1/2\boldsymbol{e}_4^{\mathrm{T}}\Big(-\boldsymbol{M}_{11}^{-1}\boldsymbol{J}_{p\theta}^{\mathrm{T}}\left(\varepsilon\boldsymbol{\tau}_1+\varepsilon^2\boldsymbol{\tau}_2\right) - \varepsilon^2\left(\boldsymbol{J}_{p\theta}^{-1}\phi_l\ddot{\boldsymbol{h}}_{10} + \dot{\overline{\boldsymbol{J}_{p\theta}^{-1}}}\phi_l\dot{\boldsymbol{h}}_{10}\right)/l_1 + \\
&\varepsilon^2\boldsymbol{M}_{11}^{-1}\left(\left(\boldsymbol{f}_1\right)_{20}\boldsymbol{h}_{10}+\left(\boldsymbol{f}_1\right)_{21}\dot{\boldsymbol{h}}_{10}\right)/2 - \varepsilon^2\boldsymbol{J}_{12}\left(\boldsymbol{J}_{22}\right)_0^{-1}\left(\boldsymbol{J}_{22}\right)_2\left(\left(\boldsymbol{f}_2\right)_0+\boldsymbol{h}_{10}\right)/2 - \varepsilon^2\left(c_1+c_2\right)\boldsymbol{J}_{p\theta}^{-1}\phi_l\dot{\boldsymbol{h}}_{10}/l_1 - \\
&\varepsilon^2\boldsymbol{J}_{12}\left(\boldsymbol{J}_{22}\right)_0^{-1}\ddot{\boldsymbol{h}}_{10} - \varepsilon^2\left(c_1c_2+1\right)\boldsymbol{J}_{p\theta}^{-1}\phi_l\boldsymbol{h}_{10}/l_1 - \varepsilon^2\boldsymbol{J}_{12}\left(f_2\right)_{21}\dot{\boldsymbol{h}}_{10}/2\Big).
\end{aligned} \tag{28}$$

Let the coefficient of ε and ε^2 be zero, and the corrective torque is,

$$\begin{aligned}
\boldsymbol{\tau}_1 &= 0, \\
\boldsymbol{\tau}_2 &= -\left(\boldsymbol{J}_{p\theta}^{\mathrm{T}}\right)^{-1}\boldsymbol{M}_{11}\Big((c_1+c_2)\left(\boldsymbol{J}_{p\theta}^{-1}\phi_l\dot{\boldsymbol{h}}_{10}\right)/l_1 + (c_1c_2+1)\boldsymbol{J}_{p\theta}^{-1}\phi_l\boldsymbol{h}_{10}/l_1 - \boldsymbol{M}_{11}^{-1}\left((\boldsymbol{f}_1)_{20}\boldsymbol{h}_{10}+(\boldsymbol{f}_1)_{21}\dot{\boldsymbol{h}}_{10}\right)/2 \\
&+ \boldsymbol{J}_{12}(\boldsymbol{J}_{22})_0^{-1}(\boldsymbol{J}_{22})_2(\boldsymbol{f}_2+\boldsymbol{h}_{10})/2 + \boldsymbol{J}_{12}(\boldsymbol{J}_{22})_0^{-1}\ddot{\boldsymbol{h}}_{10} + \left(\boldsymbol{J}_{p\theta}^{-1}\phi_l\ddot{\boldsymbol{h}}_{10} + \dot{\overline{\boldsymbol{J}_{p\theta}^{-1}}}\phi_l\dot{\boldsymbol{h}}_{10}\right)/l_1 + \boldsymbol{J}_{12}(\boldsymbol{f}_2)_{21}\dot{\boldsymbol{h}}_{10}/2\Big)
\end{aligned} \tag{29}$$

At this time, $\dot{\boldsymbol{V}} = -\boldsymbol{e}_4^{\mathrm{T}}(c_1+c_2)\boldsymbol{e}_4 \leq 0$ is valid, and the system is stable, which means the elastic displacement compensation for the end-effector's pose is realized by designing the corrective torque.

5. Sliding mode variable structure-based fast subsystem control

Define a new time scale $t_f = t/\varepsilon$, and the fast subsystem differential Eq. (12) can be expressed as,

$$\begin{cases}
\dfrac{\mathrm{d}\boldsymbol{X}_{f1}}{\mathrm{d}t_f} = \boldsymbol{X}_{f2}, \\
\dfrac{\mathrm{d}\boldsymbol{X}_{f2}}{\mathrm{d}t_f} = \boldsymbol{J}_{12}^{\mathrm{T}}\boldsymbol{J}_{p\theta}^{\mathrm{T}}\boldsymbol{\tau}_f - (\boldsymbol{J}_{22})_0\boldsymbol{X}_{f1} - \varepsilon^2\left((\boldsymbol{J}_{22})_2 + \boldsymbol{J}_{12}^{\mathrm{T}}(\boldsymbol{f}_1)_{20}\right)\boldsymbol{X}_{f1}/2 - \varepsilon\boldsymbol{J}_{12}^{\mathrm{T}}(\boldsymbol{f}_1)_{21}\boldsymbol{X}_{f2}/2
\end{cases} \tag{30}$$

The latter two terms of the second equation contain small parameter ε, and the control amount is small compared to other terms, which can be regarded as the disturbance, so the disturbance term can be expressed as,

$$\boldsymbol{\Delta}_1 = \varepsilon^2\left((\boldsymbol{J}_{22})_2 + \boldsymbol{J}_{12}^{\mathrm{T}}(\boldsymbol{f}_1)_{20}\right)\boldsymbol{X}_{f1}/2 - \varepsilon\left(\boldsymbol{J}_{12}^{\mathrm{T}}(\boldsymbol{f}_1)_{21} + (\boldsymbol{J}_{22})_0(\boldsymbol{f}_2)_{21}\right)\boldsymbol{X}_{f2}/2. \tag{31}$$

Due to the existence of the disturbance term, the fast subsystem adopts sliding mode variable structure control, and the sliding mode surface is selected as,

$$\boldsymbol{S}(t) = \boldsymbol{K}_f\boldsymbol{X}_{f1} + \boldsymbol{X}_{f2}. \tag{32}$$

where $\boldsymbol{K}_1$ is a positive number, the derivation of the upper sliding surface can be obtained,

$$\dot{S}(t) = \boldsymbol{K}_f\boldsymbol{X} + \boldsymbol{J}_{12}^{\mathrm{T}}\boldsymbol{J}_{p\theta}^{\mathrm{T}}\boldsymbol{\tau}_f - (J_{22})_0\boldsymbol{X}_{f1} - \boldsymbol{\Delta}_1. \tag{33}$$

According to the sliding surface, the Lyapunov function is defined as,

$$\boldsymbol{V}_4 = 1/2\boldsymbol{S}^{\mathrm{T}}\boldsymbol{S}. \tag{34}$$

Derivative of the above equation with respect to time can be obtained as,

$$\dot{\boldsymbol{V}}_4 = \boldsymbol{S}^{\mathrm{T}}\dot{\boldsymbol{S}} = \boldsymbol{S}^{\mathrm{T}}\left(\boldsymbol{K}_f\boldsymbol{X}_{f2} + \boldsymbol{J}_{12}^{\mathrm{T}}\boldsymbol{J}_{p\theta}^{\mathrm{T}}\boldsymbol{\tau}_f - (J_{22})_0\boldsymbol{X}_{f1} - \boldsymbol{\Delta}_1\right) \tag{35}$$

According to Eq. (35), the fast subsystem control law designed as,

$$\boldsymbol{\tau}_f = \left(\boldsymbol{J}_{12}^{\mathrm{T}}\boldsymbol{J}_{p\theta}^{\mathrm{T}}\right)^{-1}\left(-\boldsymbol{K}_f\boldsymbol{X}_{f2} + (J_{22})_0\boldsymbol{X}_{f1} - \boldsymbol{K}_f\boldsymbol{S} + \boldsymbol{\Delta}_1 \operatorname{sgn}(\boldsymbol{S})\right). \tag{36}$$

where $\operatorname{sgn}(\cdot)$ is the sign function, substituting Eq. (36) into (35) can be obtained,

$$\dot{\boldsymbol{V}}_4 = \boldsymbol{S}^{\mathrm{T}}\left(\boldsymbol{K}_f\boldsymbol{X} + \boldsymbol{J}_{12}^{\mathrm{T}}\boldsymbol{J}_{p\theta}^{\mathrm{T}}\boldsymbol{\tau}_f - (J_{22})_0\boldsymbol{X}_{f1} - \boldsymbol{\Delta}_1\right) = -\boldsymbol{\Delta}_1|\boldsymbol{S}| - \boldsymbol{\Delta}_1\boldsymbol{S} - \boldsymbol{S}^{\mathrm{T}}\boldsymbol{K}_f\boldsymbol{S} \le -\boldsymbol{S}^{\mathrm{T}}\boldsymbol{K}_f\boldsymbol{S} \le 0 \tag{37}$$

Therefore, according to the Lyapunov stability principle, the fast subsystem is convergent with torque of (36). The symbolic function will cause jitter to the system. To reduce the generation of jitter, the saturation function $sat(\cdot)$ is substituted for the symbol function. The saturation function can be defined as [20],

$$sat(s_1) = \begin{cases} 1, & s_1 > \Delta_2; \\ s_1/\Delta_2, & |s_1| \le \Delta_2; \\ -1, & s_1 < -\Delta_2. \end{cases} \tag{38}$$

where Δ_2 is the buffer layer.

6. The high-gain observer for the curvature change rate

The curvature can be obtained by strain gage measurement of the stress of the links, and the change rate of curvature is directly related to the rate of change of stress, and generally cannot be directly measured. In order to avoid direct measurement of the change rate of curvature, this chapter will design a high-gain observer to observe the curvature change rate by measuring the curvature. It can be known from Eq. (11) that the fast subsystem variable $\boldsymbol{X}_{f1}$ corresponds to the curvature, which can be directly converted by measuring the stress. $\boldsymbol{X}_{f2}$ corresponds to observed curvature change rate. According to the literature [21, 22] and the formula (4), the observer can be expressed as,

$$\begin{cases} \varepsilon\dot{\hat{X}}_{f1} = \hat{X}_{f2} + \frac{1}{\varepsilon_1} H_p \left(X_{f1} - \hat{X}_{f1} \right), \\ \varepsilon\dot{\hat{X}}_{f2} = \frac{1}{\varepsilon_1^2} H_v \left(X_{f1} - \hat{X}_{f1} \right). \end{cases} \tag{39}$$

where $\hat{X}_{f1}$ and $\hat{X}_{f2}$ represent the estimated values of X_{f1} and X_{f2}, respectively, ε_1 is the minimum positive number, H_p and H_v are the constant matrix, the observer tracking error is defined as,

$$\begin{cases} \tilde{X}_{f1} = \hat{X}_{f1} - X_{f1}, \\ \tilde{X}_{f2} = \hat{X}_{f2} - X_{f2}. \end{cases} \tag{40}$$

To prove the stability of the system, new variables of error are defined as,

$$\begin{cases} \tilde{Z}_{f1} = \tilde{X}_{f1}, \\ \tilde{Z}_{f2} = \varepsilon_1 \tilde{X}_{f2}. \end{cases} \tag{41}$$

Substitute the above equation into (39), the state observer can be expressed as,

$$\begin{cases} \varepsilon\varepsilon_1 \dot{\tilde{Z}}_{f1} = \tilde{Z}_{f2} - H_p \tilde{Z}_{f1}, \\ \varepsilon\varepsilon_1 \dot{\tilde{Z}}_{f2} = -H_v \tilde{Z}_{f1} + \varepsilon\varepsilon_1^2 \left(J_{12}^{\mathrm{T}} J_{p\theta}^{\mathrm{T}} \tau_f - (J_{22})_0 X_{f1} - \Delta_1 \right) \end{cases} \tag{42}$$

The Eq. (42) can be rewritten as,

$$\varepsilon\varepsilon_1 \dot{\tilde{Z}}_f = A_0 \tilde{Z}_f + \varepsilon\varepsilon_1^2 B_0 \left(J_{12}^{\mathrm{T}} J_{p\theta}^{\mathrm{T}} \tau_f - (J_{22})_0 X_{f1} - \Delta_1 \right). \tag{43}$$

where $A_0 = \begin{bmatrix} -H_p & I_{3\times3} \\ -H_v & 0_{3\times3} \end{bmatrix}$ and $B_0 = \begin{bmatrix} 0_{3\times3} \\ I_{3\times3} \end{bmatrix}$. All eigenvalues of A_0 can be guaranteed negative by selecting H_p and H_v, which means that A_0 is the Hurwitz matrix. Define a new Lyapunov function as,

$$V_6 = \tilde{Z}_f^{\mathrm{T}} P_1 \tilde{Z}_f. \tag{44}$$

where P_1 is the positive definite symmetry matrix, the derivation is expressed as,

$$\dot{V}_6 = \frac{1}{\varepsilon\varepsilon_1} \left(\tilde{Z}_f^{\mathrm{T}} \left(A_0^{\mathrm{T}} P_1 + P_1 A_0 \right) \tilde{Z}_f + 2\varepsilon\varepsilon_1^2 \cdot \left(J_{12}^{\mathrm{T}} J_{p\theta}^{\mathrm{T}} \tau_f - (J_{22})_0 X_{f1} - \Delta_1 \right)^{\mathrm{T}} B_0^{\mathrm{T}} P_1 \tilde{Z}_f \right). \tag{45}$$

Since A_0 is a Hurwitz matrix, there is a positive definite matrix P_1, which makes,

$$A_0^{\mathrm{T}}P_1 + P_1A_0 = -I_{3\times 3}. \tag{46}$$

$\dot{V}_6$ can be rewritten as,

$$\dot{V}_6 \le -\frac{1}{\varepsilon\varepsilon_1}\|\tilde{Z}_f\|^2 + 2\varepsilon_1\left\|\left(J_{12}^{\mathrm{T}}J_{p\theta}^{\mathrm{T}}\tau_f - (J_{22})_0X_{f1} - \Delta_1\right)^{\mathrm{T}}B_0^{\mathrm{T}}P_1\right\|\|\tilde{Z}_f\|. \tag{47}$$

According to Eq. (47), when ε_1^2 satisfied the following relationship, $\dot{V}_6 \le 0$ is established, which means the high-gain observer gradually converges,

$$\varepsilon_1^2 \le \frac{2\left\|\left(J_{12}^{\mathrm{T}}J_{p\theta}^{\mathrm{T}}\tau_f - (J_{22})_0X_{f1} - \Delta_1\right)^{\mathrm{T}}B_0^{\mathrm{T}}P_1\right\|}{\varepsilon\|\tilde{Z}_f\|}. \tag{48}$$

Therefore, according to Eq. (48), the upper bound of the small parameter can be obtained, and the fast subsystem torque can be expressed as,

$$\tau_f = \left(J_{12}^{\mathrm{T}}J_{p\theta}^{\mathrm{T}}\right)^{-1}\left(-K_f\hat{X}_{f2} + (J_{22})_0\hat{X}_{f1} - K_f\hat{S} + \Delta_1 sat\left(\hat{S}\right)\right). \tag{49}$$

where $\hat{S} = K_f\hat{X}_{f1} + \hat{X}_{f2}$. According to Eq. (12) and (42), the error equation of the fast subsystem can be expressed as,

$$\varepsilon\dot{\xi} = A_\xi\xi + h_\xi \tag{50}$$

where

$$\xi = \left[X_f\ \tilde{Z}_f\right]^{\mathrm{T}},\ X_f = \left[X_{f1}\ X_{f2}\right]^{\mathrm{T}},\ A_\xi = \begin{bmatrix} A_{\xi 11} & A_{\xi 12} \\ 0 & A_0/\varepsilon_1 \end{bmatrix}, A_{\xi 11} = \begin{bmatrix} 0_{3\times 3} & I_{3\times 3} \\ -K_f^2 & -2K_f \end{bmatrix},$$

$$A_{\xi 12} = \begin{bmatrix} 0_{3\times 3} & 0_{3\times 3} \\ (J_{22})_0 - K_f^2 & -2K_f \end{bmatrix},\ h_\xi = \begin{bmatrix} \Delta_1 sat\left(\hat{S}\right) - \Delta_1 \\ \varepsilon\varepsilon_1 B_0\left(J_{12}^{\mathrm{T}}J_{p\theta}^{\mathrm{T}}\tau_f - (J_{22})_0X_{f1} - \Delta_1\right) \end{bmatrix}.$$

According to Eq. (50), the Lyapunov function can be defined as:

$$V_5 = \varepsilon\xi^{\mathrm{T}}P_\xi\xi \tag{51}$$

where P_ξ is the symmetric positive definite matrix, Eq. (51) is derived as,

$$V_5 = \varepsilon\xi^{\mathrm{T}}\left(A_\xi^{\mathrm{T}}P_\xi + P_\xi^{\mathrm{T}}A_\xi\right)\xi + 2h_\xi^{\mathrm{T}}P_\xi\xi + \varepsilon\xi^{\mathrm{T}}\dot{P}_\xi\xi. \tag{52}$$

Since $A_{\xi 11}$ and A_0 are Hurwitz matrix, for a given symmetric positive definite matrix S_ξ, there is a symmetric positive definite matrix P_ξ that satisfies the following conditions,

$$A_\xi^{\mathrm{T}}P_\xi + P_\xi^{\mathrm{T}}A_\xi = -S_\xi. \tag{53}$$

According to the Rayleigh-Ritz inequality,

$$-\boldsymbol{\xi}^{\mathrm{T}}\boldsymbol{S}_{\xi}\boldsymbol{\xi} \leq -\lambda_{\min}(\boldsymbol{S}_{\xi})\|\boldsymbol{\xi}\|^{2}, \tag{54}$$

$$\left\|\boldsymbol{h}_{\xi}^{\mathrm{T}}\boldsymbol{P}_{\xi}\boldsymbol{\xi}\right\| \leq (\chi_{0}+\chi_{1}\varepsilon_{1})\|\boldsymbol{\xi}\|, \left\|\dot{\boldsymbol{P}}_{\xi}\right\| \leq \chi_{2}. \tag{55}$$

where $\lambda_{\min}(\cdot)$ represents the minimum eigenvalues of the corresponding matrix. χ_0, χ_1, and χ_2 are positive real numbers. According to Eqs. (53) through (55), Eq. (52) can be expressed as,

$$\dot{V}_{5} \leq -\lambda_{\min}(\boldsymbol{S}_{\xi})\|\boldsymbol{\xi}\|^{2} + \varepsilon\chi_{2}\|\boldsymbol{\xi}\|^{2} + 2(\chi_{0}+\chi_{1}\varepsilon_{1})\|\boldsymbol{\xi}\|. \tag{56}$$

According to Eq. (56), when $\dot{V}_5 \leq 0$, the small parameters in the high-gain observer satisfied $0 \leq \varepsilon_1 \leq \varepsilon_{1\max}$, the fast subsystem based on the high-gain observer is stable, and the upper bound of the small parameter satisfies the following requirements,

$$\varepsilon_{1\max} \leq (\lambda_{\min}(\boldsymbol{S}_{\xi})\|\boldsymbol{\xi}\| - \varepsilon\chi_{2}\|\boldsymbol{\xi}\| - 2\chi_{0})/\chi_{1}. \tag{57}$$

7. Stability proof of the system

The abovementioned integral manifold is used to reduce the rigid-flexible coupling system of high-speed PM, and the complex high-order system is decomposed into a slow subsystem describing the rigid body motion and a fast subsystem of elastic deformation, and the backstepping control and sliding mode variable structure control are adopted for two subsystems, respectively, and designed a high-gain observer to solve the problem that the elastic displacement change rate is difficult to measure, and proved the stability of each subsystem. However, the stability of each subsystem does not guarantee the stability of the overall system. Therefore, it is necessary to synthesize the subsystems to prove the stability of the overall system. Substituting Eqs. (9), (20), and (29) into kinetic Eq. (3), the systematic error equation can obtained,

$$\dot{\boldsymbol{e}}_{s} = \boldsymbol{A}_{s}\boldsymbol{e}_{s} + \boldsymbol{h}_{s}, \varepsilon\dot{\boldsymbol{\xi}} = \boldsymbol{A}_{\xi}\boldsymbol{\xi} + \boldsymbol{h}_{\xi}. \tag{58}$$

where

$$\boldsymbol{e}_{s} = \left[\boldsymbol{X}_{1}-\boldsymbol{X}_{d}\ \dot{\boldsymbol{X}}_{1}-\dot{\boldsymbol{X}}_{d}\right]^{\mathrm{T}}, \boldsymbol{h}_{s} = \begin{bmatrix} 0 \\ \boldsymbol{h}_{s1} \end{bmatrix}, \boldsymbol{A}_{s} = \begin{bmatrix} \boldsymbol{0}_{3\times3} & \boldsymbol{I}_{3\times3} \\ -(c_{1}c_{2}+1)\boldsymbol{I}_{3\times3} & -(c_{1}+c_{2})\boldsymbol{I}_{3\times3} \end{bmatrix},$$

$$\boldsymbol{h}_{s1} = \boldsymbol{J}_{11}\boldsymbol{J}_{p\theta}^{\mathrm{T}}\boldsymbol{\tau}_{f} - \boldsymbol{J}_{12}\boldsymbol{X}_{f1} - \varepsilon^{2}\boldsymbol{J}_{11}\boldsymbol{M}_{11}\left((c_{1}+c_{2})\boldsymbol{J}_{p\theta}^{-1}\phi_{l}\dot{\boldsymbol{h}}_{10}/l_{1} + (c_{1}c_{2}+1)\boldsymbol{J}_{p\theta}^{-1}\phi_{l}\boldsymbol{h}_{10}/l_{1}\right.$$
$$\left. + \left(\boldsymbol{J}_{p\theta}^{-1}\phi_{l}\ddot{\boldsymbol{h}}_{10} + \dot{\overline{\boldsymbol{J}_{p\theta}^{-1}}}\phi_{l}\dot{\boldsymbol{h}}_{10}\right)/l_{1}\right).$$

According to the error equation, define the Lyapunov function that contains the overall system as,

$$V_{6} = \boldsymbol{e}_{s}^{\mathrm{T}}\boldsymbol{P}_{s}\boldsymbol{e}_{s} + \varepsilon\boldsymbol{\xi}^{\mathrm{T}}\boldsymbol{P}_{\xi}\boldsymbol{\xi}. \tag{59}$$

where P_s and P_ξ are the symmetric positive definite matrix, the derivative of Eq. (59) can be obtained,

$$\dot{V}_6 = e_s^{\mathrm{T}}\left(A_s^{\mathrm{T}}P_s + P_s^{\mathrm{T}}A_s\right)e_s + \xi^{\mathrm{T}}\left(A_\xi^{\mathrm{T}}P_\xi + P_\xi^{\mathrm{T}}A_\xi\right)\xi + 2h_s^{\mathrm{T}}P_s e_s + 2h_\xi^{\mathrm{T}}P_\xi\xi + \varepsilon\xi^{\mathrm{T}}\dot{P}_\xi\xi. \tag{60}$$

Since A_s is a Hurwitz matrix, for a given symmetric positive definite matrix S_s, there is a symmetric positive definite matrix P_s that satisfies the following conditions,

$$A_s^{\mathrm{T}}P_s + P_s^{\mathrm{T}}A_s = -S_s. \tag{61}$$

According to Eqs. (53) and (61), $\dot{V}_6$ can be rewritten as,

$$\dot{V}_6 = -e_s^{\mathrm{T}}S_s e_s - \xi^{\mathrm{T}}S_\xi\xi + 2h_s^{\mathrm{T}}P_s e_s + 2h_\xi^{\mathrm{T}}P_\xi\xi + \varepsilon\xi^{\mathrm{T}}\dot{P}_\xi\xi \tag{62}$$

According to the Rayleigh-Ritz inequality, we can obtain,

$$-e_s^{\mathrm{T}}S_s e_s \le -\lambda_{\min}(S_s)\|e_s\|^2, \tag{63}$$

$$-\xi^{\mathrm{T}}S_\xi\xi \le -\lambda_{\min}(S_\xi)\|\xi\|^2, \tag{64}$$

$$\left\|h_s^{\mathrm{T}}P_s e_s\right\| \le \left(\chi_3 + \chi_4\varepsilon + \chi_5\varepsilon^2\right)\|e_s\|\|\xi\|, \tag{65}$$

$$\left\|h_\xi^{\mathrm{T}}P_\xi\xi\right\| \le \left(\chi_6 + \chi_7\varepsilon + \chi_8\varepsilon^2\right)\|\xi\|^2. \tag{66}$$

where $\chi_i(i = 0, 1, \ldots 6)$ is positive. According to the inequality relationship shown by Eqs. (63) to (66), $\dot{V}_6$ satisfied the following relationship,

$$\dot{V}_6 \le -\left[\|e_s\| \quad \|\xi\|\right]\cdot\begin{bmatrix}\lambda_{\min}(S_s) & -\left(\chi_3+\chi_4\varepsilon+\chi_5\varepsilon^2\right) \\ -\left(\chi_3+\chi_4\varepsilon+\chi_5\varepsilon^2\right) & \lambda_{\min}(S_\xi) - 2\left(\chi_6+\chi_7\varepsilon+\chi_8\varepsilon^2\right) - \chi_2\varepsilon\end{bmatrix}\begin{bmatrix}\|e_s\| \\ \|\xi\|\end{bmatrix}. \tag{67}$$

The condition that the closed-loop system is asymptotically stable is $\dot{V}_6 \le 0$, from the above equation, the condition of $\dot{V}_6 \le 0$ is that the coefficient matrix is positive, that is,

$$\lambda_{\min}(S_s)\left(\lambda_{\min}(S_\xi) - 2\left(\chi_6+\chi_7\varepsilon+\chi_8\varepsilon^2\right) - \chi_2\varepsilon\right) - \left(\chi_3+\chi_4\varepsilon+\chi_5\varepsilon^2\right)^2 \ge 0. \tag{68}$$

Ignoring the influence of high-order terms of $O(\varepsilon^2)$, when the maximum value of the small parameter ε satisfied,

$$\varepsilon_{\max} = \frac{-\lambda_b + \sqrt{\lambda_b^2 + 4\lambda_a\lambda_c}}{2\lambda_a}. \tag{69}$$

$\dot{V}_6 \le 0$ is valid, where

$$\begin{cases} \lambda_a = \lambda_{\min}(S_s)\chi_8 + \chi_4^2 + 2\chi_3\chi_5, \\ \lambda_b = -2\lambda_{\min}(S_s)\chi_7 - \lambda_{\min}(S_s)\chi_2 - 2\chi_3\chi_4, \\ \lambda_c = \lambda_{\min}(S_s)\lambda_{\min}(S_\xi) - 2\lambda_{\min}(S_s)\chi_6 - \chi_3^2. \end{cases} \tag{70}$$

According to Eq. (67), when the value of ε satisfied $0 < \varepsilon \leq \varepsilon_{max}$, the overall system is stable.

8. Algorithm simulations

When the Taylor expanding order $p = 0$ is valid, the integral manifold (IM) is equivalent to the singular perturbation (SP). In order to verify the composite control proposed in this chapter, this section compares it with the singular perturbation control and the backstepping (BS) control considering only the rigid-body dynamic model. The above algorithm simulation will be carried out under the SIMULINK module of the MATLAB software, and the ode15s integral will be selected. According to formula (29), in the composite control algorithm based on the integral manifold and observer, the desired trajectory of the end-effector of the moving platform needs to satisfy the fourth derivative continuous, and at the same time to reduce the impact to the system at the beginning and end point of the desired trajectory. The nine-order polynomial shown in Eq. (71) is used to ensure that the velocity, acceleration, and the third and fourth derivatives at the start and end points are zero.

$$\begin{cases} p_x = A_0\left(125t^5/t_d^5 - 420t^6/t_d^6 + 540t^7/t_d^7 - 315t^8/t_d^8 + 70t^9/t_d^9\right) + p_{x0}, \\ p_y = p_{y0}, \\ \phi = 0. \end{cases} \tag{71}$$

where the running time t_d is 0.06 s, the starting position $p_{x0} = 187.5$, $p_{y0} = 187.5/\sqrt{3}$, and the amplitude $A_0 = 30$ of the desired trajectory. Take $\varepsilon^2 = 1/k_s$, $\Delta_1 = 1 \times 10^{-3}$, $c_1 = c_2 = 50$, $\Delta_2 = 0.05$, $\boldsymbol{H}_p = \mathbf{diag}([40, 40, 40])$, $\boldsymbol{H}_v = \mathbf{diag}([400, 400, 400])$, $\boldsymbol{K}_f = \mathbf{diag}([60, 60, 60])$. According to Eq. (57), take $\varepsilon_1 = 0.001$. The parameters added and modified in [17] are as follows: the height and thickness of the links are 30 and 5 mm, respectively, the reduction ratio is 20, and the moment of inertia between the motor and the reducer is 284.1 $kg \cdot mm^2$.

To describe the control performance of the end-effector, an average error is introduced, and is defined as,

$$\begin{cases} t_M = \sqrt{\dfrac{1}{t_d}\displaystyle\int_0^{t_d}\left(C_R(1)^2 + C_R(2)^2\right)\mathrm{dt}} \\ r_M = \sqrt{\dfrac{1}{t_d}\displaystyle\int_0^{t_d} C_R(3)^2\mathrm{dt}} \end{cases} \tag{72}$$

where C_R represents the performance index of the three directions of the moving platform, t_{rM} and r_M are the average error of the translation direction and the rotation direction.

According to Eq. (24), the elastic displacement f_3 of the moving platform can be calculated. v_i and v_m represent the maximum elastic displacement and the average elastic displacement in all directions of the moving platform during operation, v_{end} indicates the elastic displacement at the end point (residual vibration). For the same expected input, the magnitude of the elastic displacement of the moving platform can reflect the vibration suppression effect of the three control algorithms. The elastic displacements in all directions are shown in **Figures 4** and **5**, which shows that the maximum elastic displacement amplitude in all directions is reduced by more than 28% compared with the backstepping control, and the composite control is reduced by 4.75, 33.42, and 33.52% compared with the singular perturbation. The average elastic displacement for the translational direction decreases from 1.579 and 1.112 mm for backstepping control and singular perturbation to 0.970 mm for composite control. For the rotational direction, 0.0014 and 9.863×10^{-4} rad from backstepping control and singular perturbation drops to 6.872×10^{-4} rad of the composite control. Compared with the above algorithm, the elastic displacement of the composite control decreases by more than 14% in both directions. Compared with the backstepping control, when the composite control and the singular perturbation algorithm are used, the residual vibration is greatly reduced, and both algorithms are close to zero.

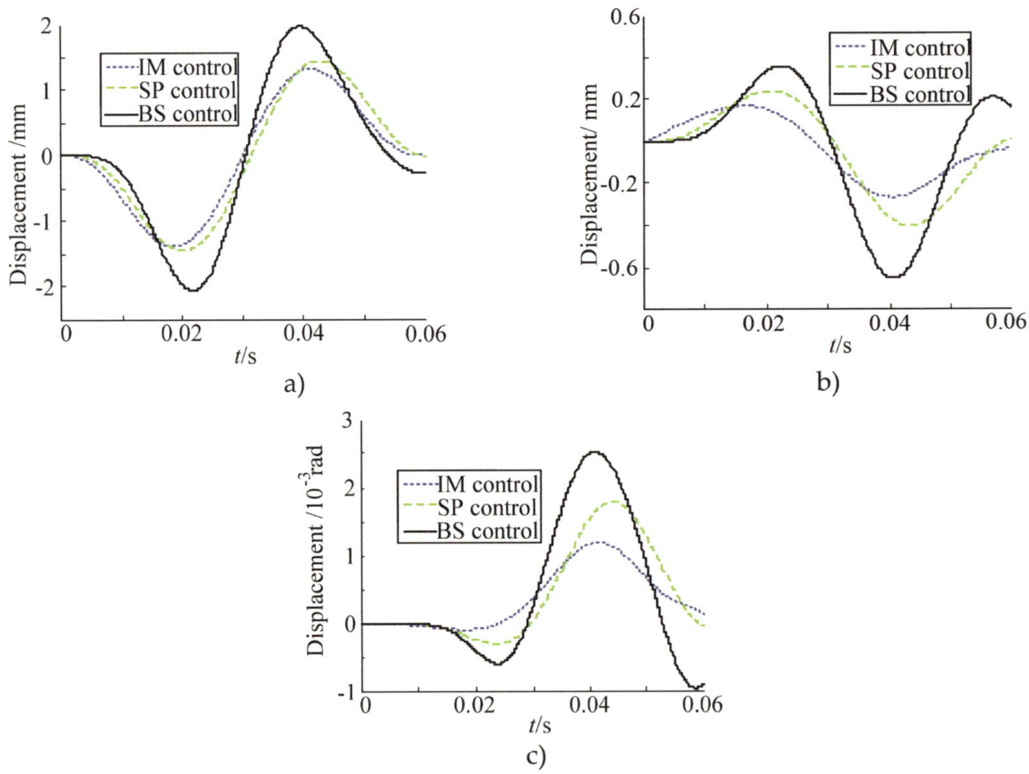

Figure 4. Flexible displacement of moving platform. (a) Displacement of X direction. (b) Displacement of Y direction. (c) Displacement of rotational direction.

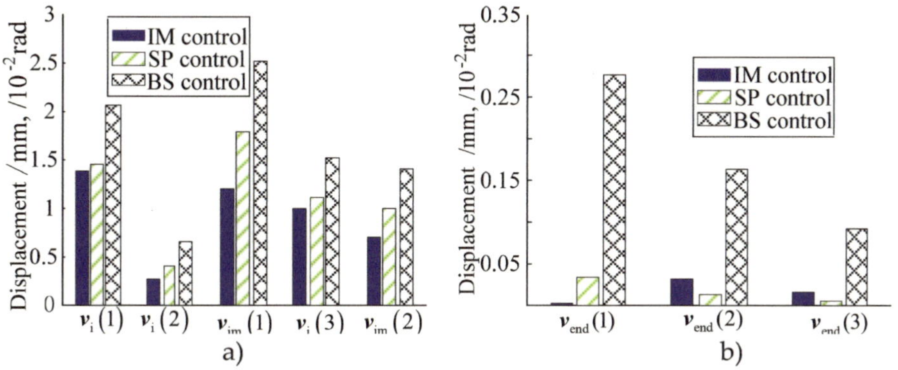

Figure 5. Vibration of the moving platform. (a) Flexible displacement of all directions. (b) Residue vibration of all directions.

The tracking error is the difference between the actual output and the desired output of the end of the moving platform. $\boldsymbol{t}_{\mathrm{r}}$ indicates the maximum tracking error in all directions of the moving platform during the whole running process, $\boldsymbol{t}_{\mathrm{rm}}$ indicates the average tracking error of the translational and rotational directions, $\boldsymbol{t}_{\mathrm{end}}$ is the tracking error at the end point. As shown in **Figures 6** and **7**, compared with the singular perturbation and backstepping control,

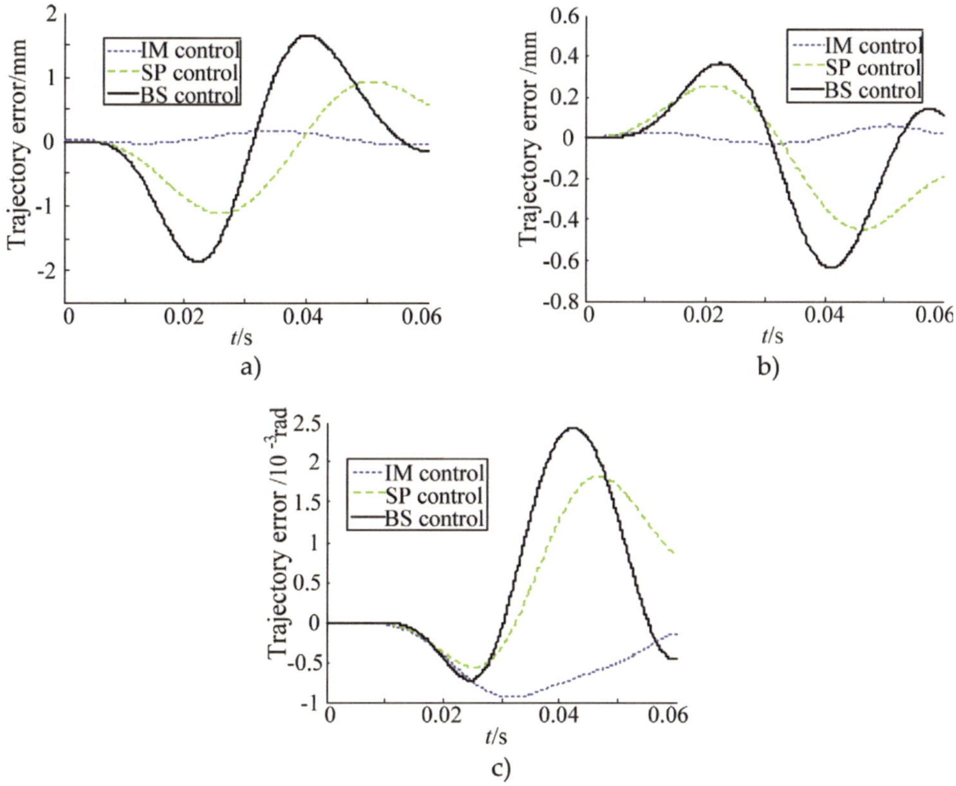

Figure 6. Trajectory error of directions. (a) Trajectory error of X direction. (b) Trajectory error of Y direction. (c) Trajectory error of rotational direction.

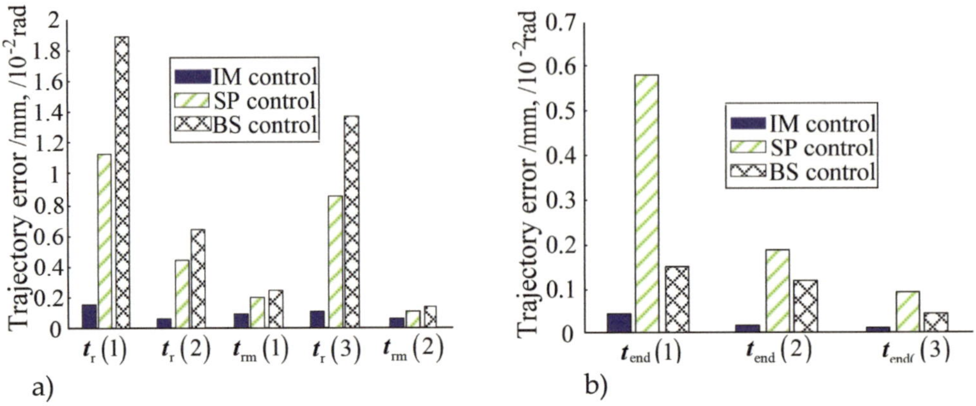

Figure 7. Tracking error of the moving platform. (a) Trajectory error of directions. (b) Residue error of directions.

the composite control based on the integral manifold and the observer has obvious advantages in trajectory tracking. For the maximum tracking error, the *X* direction decreased by 85.56 and 91.41%, and the *Y* direction decreased by 57.55 and 90.57%, while the rotation direction decreased by 53.34 and 61.5%, respectively. For the average tracking error, the translation direction decreased by 88.2 and 92.62%, the rotational direction decreased by 37.26 and 49.57%, respectively; in the tracking error of the end point, the *X* direction decreased by 92.8 and 72.34%, and the Y direction decreased by 89.73 and 83.62%, respectively, while the rotational direction decreased by 85.96 and 70.85%, respectively. For the tracking error at the end point, the singular perturbation method is significantly worse than the backstepping controller in all directions. This is mainly because the singular perturbation algorithm only considers the vibration suppression, and the cost of the vibration suppression is at the cost of sacrificing the trajectory tracking due to the delay of the adjustment. It can be seen from the above analysis that in the aspect of trajectory tracking accuracy, the composite control based on integral manifold and observer has significant advantages.

9. Conclusions

1. Decompose the rigid-flexible coupling dynamic model into fast and slow subsystems based on the integral manifold, and employ the sliding mode control and backstepping control to design the fast and slow subsystem controllers, respectively, and compensate the elastic displacement at the end of the manipulator. A high-gain observer estimates the rate of change of curvature, which in turn enables trajectory tracking control of high-speed PM.

2. The Lyapunov function is selected to prove the asymptotic stability of the slow subsystem, fast subsystem, high-gain observer, and the overall system. The conditions for selecting the integral manifold and the small parameters of the observer are given.

3. Apply MATLAB-SIMULINK to establish a comparison simulation to verify the performance of the proposed compound control algorithm. The simulation results show that the composite control algorithm has obvious advantages in vibration suppression and trajectory tracking.

Acknowledgements

This research was supported in part by the Natural Science Foundation of Zhejiang under Grant No. LY18E050019 and the Excellent Talent Cultivation Foundation under Grant No. ZSTUME02B09.

Author details

Zhengsheng Chen[1,2*]

*Address all correspondence to: zschen88200@163.com

1 School of Mechatronics Engineering, Harbin Institute of Technology, Harbin, China

2 Department of Mechatronics Engineering, Zhejiang Sci-tech University, Hangzhou, China

References

[1] Pietsch I, Krefft M, Becker O, et al. How to reach the dynamic limits of parallel robots? An autonomous control approach. IEEE Transactions on Automation Science and Engineering. 2005;**2**(4):369-380

[2] Dwivedy S, Eberhard P. Dynamic analysis of flexible manipulators, a literature review. Mechanism and Machine Theory. 2006;**41**(7):749-777

[3] Yongming LI, Tong S, Tieshan LI. Adaptive fuzzy output feedback control for a single-link flexible robot manipulator driven DC motor via backstepping. Nonlinear Analysis: Real World Applications. 2013;**14**(1):483-494

[4] Vakil M, Fotouhi R, Nikiforuk P. Causal end-effector inversion of a flexible link manipulator. Mechatronics. 2009;**19**(7):1197-1210

[5] Moallem M, Patel R, Khorasani K. Nonlinear tip-position tracking control of a flexible-link manipulator, theory and experiments. Automatica. 2001;**37**(11):1825-1834

[6] Lizarraga I, Etxebarria V. Combined PD-H_∞ approach to control of flexible link manipulators using only directly measurable variables. Cybernetics and Systems. 2003;**34**(1):19-31

[7] Zhang Q, Mills J, Cleghorn W, et al. Trajectory tracking and vibration suppression of a 3-PRR parallel manipulator with flexible links. Multibody System Dynamics. 2015;**33**(1): 27-60

[8] El-Badawy A, Mehrez MW, Ali AR. Nonlinear modeling and control of flexible-link manipulators subjected to parametric excitation. Nonlinear Dynamics. 2010;**62**(4):769-779

[9] Subudhi B, Morris A. Soft computing methods applied to the control of a flexible robot manipulator. Applied Soft Computing. 2009;**9**(1):149-158

[10] Khorasani K. Adaptive control of flexible-joint robots. IEEE Transactions on Robotics and Automation. 1992;**8**(2):250-267

[11] Moallem M, Khorasani K, Patel R. An integral manifold approach for tip-position tracking of flexible multi-link manipulators. IEEE Transactions on Robotics and Automation. 1997; **13**(6):823-837

[12] Salmasi H, Fotouhi R, Nikiforuk P. A manoeuvre control strategy for flexible-joint manipulators with joint dry friction. Robotica. 2010;**28**(04):621-635

[13] Vakil M, Fotouhi R, Nikiforuk P. Application of the integral manifold concept for the end-effector trajectory tracking of a flexible link manipulator. In: Proceedings of the 26th American Control Conference. New York: IEEE; 2007. pp. 741-747

[14] Vakil M, Fotouhi R, Nikiforuk P. End-effector trajectory tracking of a class of flexible link manipulators. In: Proceedings of 32nd Annual Mechanisms and Robotics Conference. New York: ASME; 2008. pp. 1085-1094

[15] Vakil M, Fotouhi R, Nikiforuk P. End-effector trajectory tracking of a flexible link manipulator using integral manifold concept. International Journal of Systems Science. 2011; **42**(12):2057-2069

[16] Vakil M, Fotouhi R, Nikiforuk P. Maneuver control of the multilink flexible manipulators. International Journal of Non-Linear Mechanics. 2009;**44**(8):831-844

[17] Chen Z, Kong M, Ji C, et al. An efficient dynamic modelling approach for high-speed planar parallel manipulator with flexible links. Proceedings of the Institution of Mechanical Engineers, Part C: Journal of Mechanical Engineering Science. 2015;**229**(4):663-678

[18] Gorius T, Seifried R, Eberhard P. Approximate end-effector tracking control of flexible multibody systems using singular perturbations. Journal of Computational and Nonlinear Dynamics. 2013;**9**(1):011017

[19] Ge Z, Lee T, Ge S. Tip tracking control of a single-link flexible robot, a backstepping approach. Dynamics and Control. 1997;**7**(4):341-360

[20] Lee S, Lee C. Hybrid control scheme for robust tracking of two-link flexible manipulator. Journal of Intelligent and Robotic Systems. 2001;**32**(4):389-410

[21] Heredia J, Wen Y. A high-gain observer-based PD control for robot manipulator. In: Proceedings of the 2000 American Control Conference. Chicago: IEEE; 2000. pp. 2518-2522

[22] Mosayebi M, Ghayour M, Sadigh M. A nonlinear high gain observer based input-output control of flexible link manipulator. Mechanics Research Communications. 2012;**45**:34-41

Chapter 6

Manifold-Based Robot Motion Generation

Yuichi Kobayashi

Additional information is available at the end of the chapter

http://dx.doi.org/10.5772/intechopen.80973

Abstract

In order to make an autonomous robot system more adaptive to human-centered environments, it is effective to let the robot collect sensor values by itself and build controller to reach a desired configuration autonomously. Multiple sensors are often available to estimate the state of the robot, but they contain two problems: (1) sensing ranges of each sensor might not overlap with each other and (2) sensor variable can contain redundancy against the original state space. Regarding the first problem, a local coordinate definition based on a sensor value and its extension to unobservable region is presented. This technique helps the robot to estimate the sensor variable outside of its observation range and to integrate regions of two sensors that do not overlap. For a solution to the second problem, a grid-based estimation of lower-dimensional subspace is presented. This estimation of manifold allows the robot to have a compact representation, and thus the proposed motion generation method can be applied to the redundant sensor system. In the case of image feature spaces with a high-dimensional sensor signal, a manifold estimation-based mapping, known as locally linear embedding (LLE), was applied to an estimation of distance between robot body and an obstacle.

Keywords: robot motion generation, redundant sensors, limited observation range, manifold by constraint

1. Introduction

Robotics is gathering attention for various applications such as autonomous navigation and manipulation of objects. It is highly expected that autonomous robots can act closer to humans, for example, in household environment. In reality, however, it is still very difficult to make those robots achieve various tasks in environments, which are not specifically structured for the robots. One of the reasons for this is that processes of recognition and motion generation are all specifically designed according to individual specific cases. In unstructured environments, robots have to adapt to various changes of conditions in both recognition and motion control processes, which requires reconstruction of software by the human designers.

One possible approach to this problem, which might be promising but not straightforward, is to let the robot learn to build its representation for task execution based on its own experience (for example, discussed in the context of developmental robotics [1, 2]). In this approach, task-specific designs are omitted in recognition processes, which is quite different from the conventional robotics, where objects, robots, and environments are described by their coordinates (typically Cartesian) in world coordinate systems. For example, a mobile robot can achieve a navigation task based only on its information of distance sensors, while distance sensor information is normally converted to position of the robot based on its environmental map in simultaneous localization and mapping (SLAM) applications [3–5].

When we try to build a framework to allow an autonomous robot to build a state space for motion generation, the idea of manifold where only local coordinate systems are defined and relations among them are described is suitable for the purpose. The reason is that one kind of sensor does not provide thorough information about the robot system and its environment, and multiple sensors are often required to cover various situations, whereas relations among multiple sensor signals are not known in advance. Thus, an application of approximating manifold for robot motion generation is presented in this chapter.

First, an integration of multiple sensor spaces is presented. The proposed integration method is based on an idea that the system dynamics is continuous over a sensor signal space with respect to the control input. Redundant sensor signals are mapped onto a lower-dimensional subspace using a simple grid-based parameterization method, which was applied to a navigation problem of a mobile robot equipped with several distance sensors measuring distances to a wall. Second, an application of locally linear embedding (LLE) [6] to mapping from a high-dimensional image feature space to a low-dimensional space in robotic motion planning task is presented. No prior knowledge on the robot appearance is used in the method, and it was shown that the obtained low-dimensional space reflected the spatial relation between the robot hand and the object.

2. Integration of multiple sensor spaces with mapping to manifold

In this section, a motion generation method using an integration of multiple sensor spaces is presented. Multiple sensors are often required to realize thorough understanding about the environment, but they often do not overlap with each other; in the case of visual recognition as an example, occlusion and restriction of the viewing range often cause an incomplete state identification. In the case of tactile and proximity sensors, their detection ranges are limited and provide useful information only in limited cases, when they are close to objects or environments. On the other hand, occlusion often occurs when the sensor is close to an object. Robots can identify their surroundings by integrating multimodal sensors, but it causes a problem of integrating information of multiple different sensors whose detection ranges do not overlap among each other.

A standard way to integrate multiple sensor information in robotics is to rely on sensor models and calibrations, but they require a preparation cost by human designers. For realizing highly adaptive autonomous robots, following properties are required:

- Integration of multimodal sensor spaces, each of which has its specific sensing range (possibly not overlapping with each other).
- Relying only on sensing and actuating information of the robot itself, without using world coordinate system models.

A motion generation framework based on multiple sensors with limited sensing ranges has been presented in [7]. In order to integrate two sensor spaces, an idea of extending a sensor space was proposed, borrowing an idea from diffusion-based learning [8, 9]. The characteristic of the proposed framework is that it can generate desired trajectory and motion without a problem-specific knowledge. It is known that the similar class of problems has been discussed using partially observable Markov decision processes (POMDPs) [10–13]. The proposed framework does not take the noise or perceptual aliasing into account, but the proposed framework is simpler.

2.1. Problem definition of motion generation with multiple sensors

Let $x \in X$ denote the state of the robot system, where $X \subset \mathbb{R}^n$ denotes the state space. Observation vectors are denoted by $s \in s^{(1)}, \dots, s^{(h)} \in s^{(h)}$, where h denotes the number of sensors and $S^{(i)} \subset \mathbb{R}^n$ denotes the observation variable space for sensor i. The control input (motor command) to the system is denoted by $u \in \mathbb{R}^n$. The dynamics of the system is expressed as

$$\dot{x} = F(x)u, F(x) \in \mathbb{R}^{n \times n}, \quad (1)$$

where $F(x)$ is a smooth function, which will be approximated locally. Each sensor' sensing range is limited. $X^{(i)} \subset X$ denotes a subset of state space where sensor i is in its sensing range. Mapping from $X^{(i)}$ to $S^{(i)}$ is assumed to be injective and smooth, where all mappings are unknown. It is also assumed that there is no noise in the observation and the robot can judge whether each sensor is in its sensing range.

The task of the robot is to move from an initial configuration x_{start} to x_{goal}, where the information of the target configuration is given as an observation vector sensed at x_{goal}. That is, the target sensor value is given to the robot as $s^{(h)}_{goal}$ where j satisfies $x_{goal} \in X^{(j)}$. As indicated in **Figure 1**, a single sensor does not cover both x_{start} and x_{goal}, nor is it guaranteed that the sensing range of one sensor does not overlap with that of another sensor. Thus, the robot must find a trajectory that goes through a subset of the state space where no sensor signal can be observed.

The second aspect of the problem is that observation variable itself contains redundancy. Let m denote the dimension of the observation variable and the redundancy means $m > n$, which

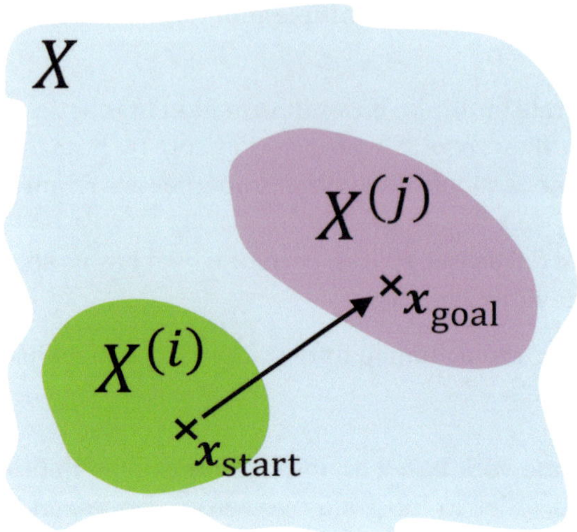

Figure 1. Motion from a sensing range of a sensor to another range.

is depicted in **Figure 2** with the case $m = 3$ and $n = 2$. The observation variables are constrained on a two-dimensional manifold. Knowing the fact of constraint on a manifold, lower dimension can be obtained by approximating the manifold.

2.2. Integration of multiple sensor spaces

The robot first acquires the mapping from the control input to an observation vector by collecting samples by random motion within each sensor's detection range. The basic idea of the integration is to first extend the mapping from outside the sensing range as indicated in **Figure 3**. The robot repeats motion in and out of the sensing range and compares the resultant observation and a predicted observation. That is, the robot estimates the observation using the information of its sequential motion and the input-observation mapping.

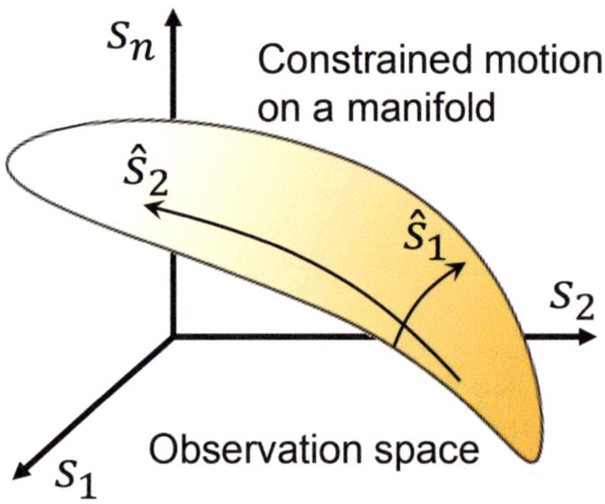

Figure 2. Redundant sensor information and constraint on a manifold.

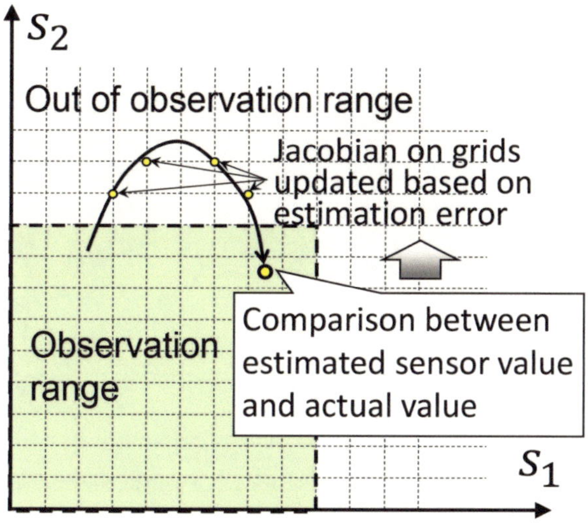

Figure 3. Estimation of Jacobian of observation variable dynamics outside observation range of a sensor.

The process of extension of a sensor space can be understood as a construction of a "virtual" observation space. The virtual observation space overlaps with another sensor space. When a task to reach a destination is given to the robot, it generates a motion from the current sensor space to the other sensor space including the destination, which is based on the representation of the virtual observation space. This framework basically works on the basis that the dimensions of the sensor spaces are equal. To relax the condition, we also discuss the way to deal with a case of redundant sensor space where an observation vector has higher dimension than the state vector.

2.3. Dimension reduction of the observation vector

Dimension reduction of sensor variable is based on a grid-based parameterization, as shown in **Figure 4**. Basic idea of the parameterization is similar to an active contour model used in image processing [14]. The nodes in a two-dimensional grid fit along the surface of samples by minimizing an energy representing closeness to the samples. By extending and fixing the nodes on the ends of the grid to the end of samples, iterative updates minimizing the energy lead the grids to fit the samples while spreading to cover the sample region. Once the lower-dimensional grid is created, it is used to parameterize the original sensor signal by another vector, in the example case, in two-dimensional vector.

2.4. Motion generation by integrating two observation spaces

Using the extrapolation of Jacobian in the observable region to outside the viewing range, virtual observation variables can be obtained. As shown in **Figure 5**, the robot starts motion from a viewing range of sensor i. The target is given as a variable of sensor j. Using the extrapolation, the closest grid in the virtual sensor space of sensor i can be calculated. First, the robot is controlled to aim at the grid in the virtual sensor space using Jacobian of sensor i. After it

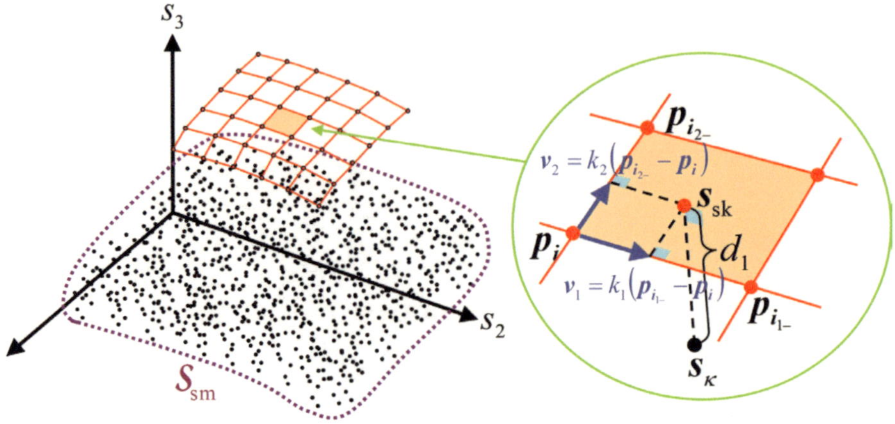

Figure 4. An example of a grid-based parameterization of sensor space with three-dimensional sensor variable. Input in the three-dimensional space is mapped onto a two-dimensional vector.

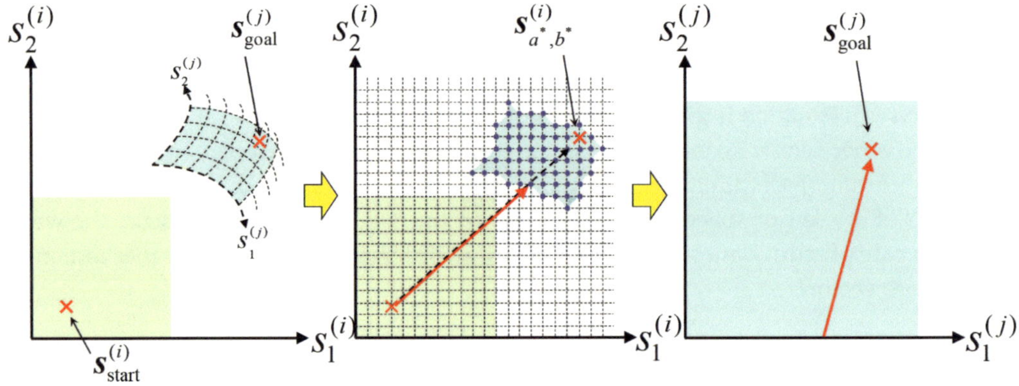

Figure 5. Motion generation by integrating two observation space, starting from a sensor space aiming at another sensor space and finally reaching the target.

reaches the viewing range of sensor j, it switches to the Jacobian of sensor j to finally reach the precise target position in the space of sensor j.

2.5. Simulation results

Consider a mobile robot with five proximity sensors, as shown in **Figure 6**. The robot is equipped with three proximity sensors on the front of its body that are grouped as sensor 1. The robot has two proximity sensors on its right side. They are grouped as sensor 2. A wall that has an infinite length is located in the environment. Each proximity sensor provides a value that is proportional to the distance to the wall. When the distance is longer than its sensing range, the wall cannot be detected. This situation is assumed to be detected by the robot. The control input to the system is the angular velocities of the two wheels.

http://dx.doi.org/10.5772/intechopen.80973

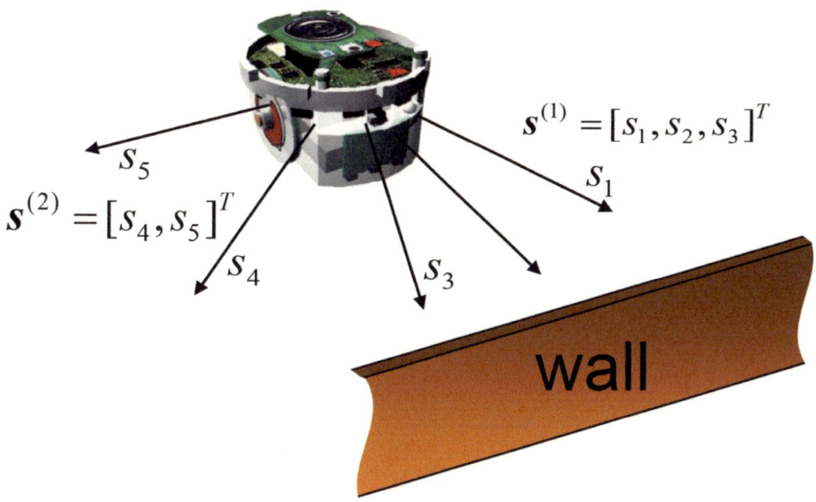

Figure 6. Mobile robot navigation problem where five distance sensors are available to detect a relative configuration against a wall.

In this problem setting, the relative position of the robot to the wall can be expressed by two parameters; its distance to the wall and its relative orientation to the wall. When the robot is moving parallel to the wall, sensor 2 detects the configuration of the robot. In this case, sensor 1 is out of its sensing range. Sensor 1 provides configuration information when the robot is facing the wall. However, in this case, the observation variables $s^{(1)} = [s_1\ s_2\ s_3]^T$ are redundant for the purpose of specifying the 2-DOF configuration relative to the wall. The mapping method of manifold is applied to the space of sensor 1. The radius of the wheels is 0.02 (m), and the distance between the two wheels is 0.04 (m). The initial state of the robot is set where the robot faces perpendicular to the wall at a distance of 1 (m), and final destination of the robot is specified so that the robot comes close to the wall where only sensor 2 is in its detection range.

Figure 7(a) shows samples of the observation variables of sensor 1, which was obtained by the offline random data collection. The three-dimensional vectors are distributing on a

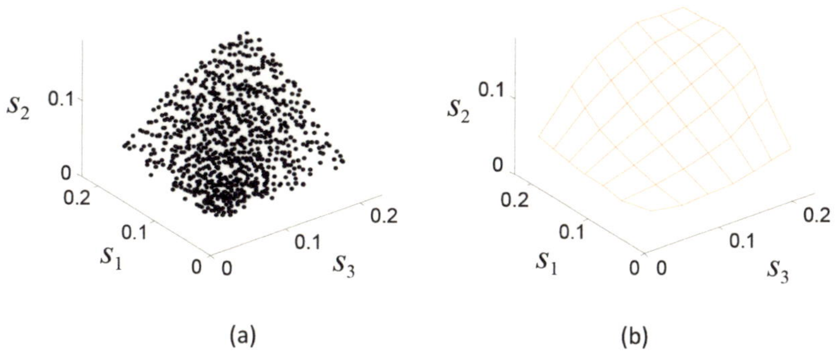

Figure 7. Collected sensor variables with three distance sensors (a) and estimated two-dimensional manifold based on the samples (b).

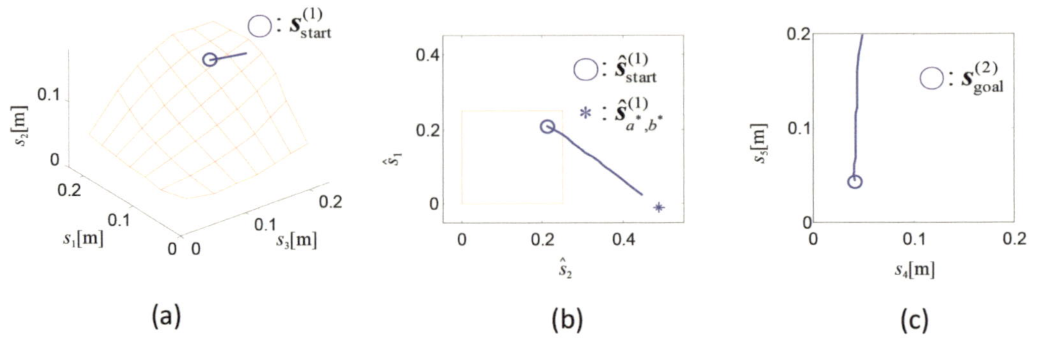

Figure 8. Realized trajectories of the robot in sensor space 1 (a), virtual sensor space of sensor 1 (b), and the viewing range of sensor 2 (c).

two-dimensional surface of the redundancy described above. Nodes of the approximation surface were initially located around the center of the samples. The approximation surface obtained by the proposed method is depicted in **Figure 7(b)**, where the nodes correspond to crosses on the curves. It can be confirmed that the nodes covered the samples by spreading and fitting them.

The trajectory obtained by the proposed method in the observation variable spaces is shown in **Figure 8**. The line drawn on the approximation surface in **Figure 8(a)** indicates the initial part of the trajectory in the observation variable space for sensor 1, where a circle in the figure indicates the initial configuration. **Figure 8(b)** shows the trajectory drawn in the space of sensor 1 obtained by the proposed dimension-reduction method. **Figure 8(c)** shows the last part of the trajectory in the space of sensor 2.

The trajectory generated in the world coordinate by the robot is depicted in **Figure 9**. The line of y = 0 indicates the wall. The initial configuration of the robot is apart from the wall and it

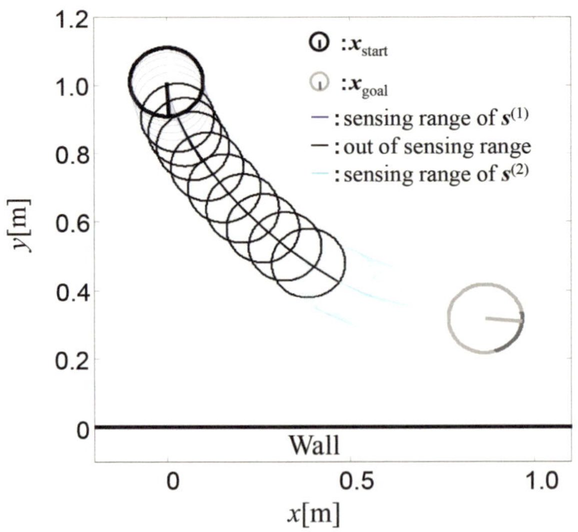

Figure 9. Trajectory of the mobile robot in the world coordinate view.

finally reached the target configuration, directing parallel to the wall. It can be seen that there is an intermediate part in the trajectory, where none of the distance sensors were detecting the wall, as drawn in thick lines in the trajectory.

It was assumed in the proposed method that the robot knows the dimension of the state vector. This problem can be resolved by applying statistical methods such as principal component analysis (PCA) [15], which allows to detect the appropriate dimension of the robot system's dynamics. Though PCA is a linear framework, which is valid only in the case where linear dimension reduction can be applied to the whole state space, nonlinear extensions of the dimension-reduction methods have been also developed, such as Kernel PCA [16] and ISOMAP [17]. The surface-approximation scheme applied in this chapter for the dimension-reduction problem can be replaced to other nonlinear mapping methods, which will be one of our future works.

3. Manifold learning approach toward constructing state representation for robot motion generation

Monocular and stereo cameras are widely used as external sensors for robot systems. In the real-world application of robot systems, however, measurement of 3D configurations of objects suffers from the following difficulties:

1. 3D configuration measurement, in general, inherently requires precise measurement of the shape of an object, but the whole shape of an object cannot be measured directly because the process is normally unilateral.
2. It is very important for object manipulation of a robot that the spatial relation between a robot and an object is precisely identified. But while the robot hand is approaching to the object and getting close to it, occlusion is very likely to occur.
3. In real applications, objects very often deform by contact with the robot, which requires specific model for mathematical analysis. But it is difficult to precisely model the deformation.

In the research field of developmental robotics, measurement of the 3D configuration in the world coordinate is not regarded as a sole way to represent the state for a robot. If a robot can build a suitable representation of its environment based by its own way, the total process of robot recognition and motion generation will be freed from the problems mentioned above (e.g., see [18] as a learning approach).

Thus, an approach to the interest of building a representation of a robot from images for motion planning and control in an adaptive way without any predefined knowledge [19] is presented in this section. To consider relation between the robot and its environment, image features based on scale invariant feature transform (SIFT) [20] are used. As a related research, an image feature-based learning of robot behavior was presented [21]. However, it did not deal with relation between an object and the robot with a quantitative representation. In the presented method, a manifold learning method is applied to acquisition of state representation. It allows not only to classify state of the robot but also to evaluate closeness to a certain situation. In addition, it is verified that the representation acquired by the method is used for motion generation of collision avoidance.

As a means of manifold learning, locally linear embedding (LLE) [6] is used. The manifold learning is suitable because the system dynamics property can hold only in a local region in the problem of robot motion generation. A vector generation based on SIFT features matching is proposed for the application of LLE to deal with the problem that keypoints of SIFT are not consistently observed throughout the image sequences. The proposed method is evaluated using a humanoid robot with real images after verification of LLE state representation generation with simulated images.

3.1. Problem definition of manifold learning from an image

Figure 10 shows images obtained by CCD camera attached at the head of a robot. These images are input to the system. Humanoid robot NAO [22] is considered in the experiment. The images contain part of the body (arm) of the robot, an object that has possibility to contact with the robot, and other objects that are not affected by the robot motion (background). Shoulder roll joint and shoulder pitch joint are controlled, while other two joints are fixed throughout the experiment. This implies that the motion of the robot arm is constrained on a plane that is vertical to optical axis of the CCD camera.

The right hand of **Figure 10** also shows image features extracted from the images as depicted by circles. Keypoints of SIFT are used as image features. No explicit knowledge on properties of image features is assumed in the problem. That is, the robot does not have label information of the object, backgrounds, or robot's body in the image in advance. The robot collects images while moving its arm randomly. Position of the object is also differed irrelevantly to the configuration of the robot arm.

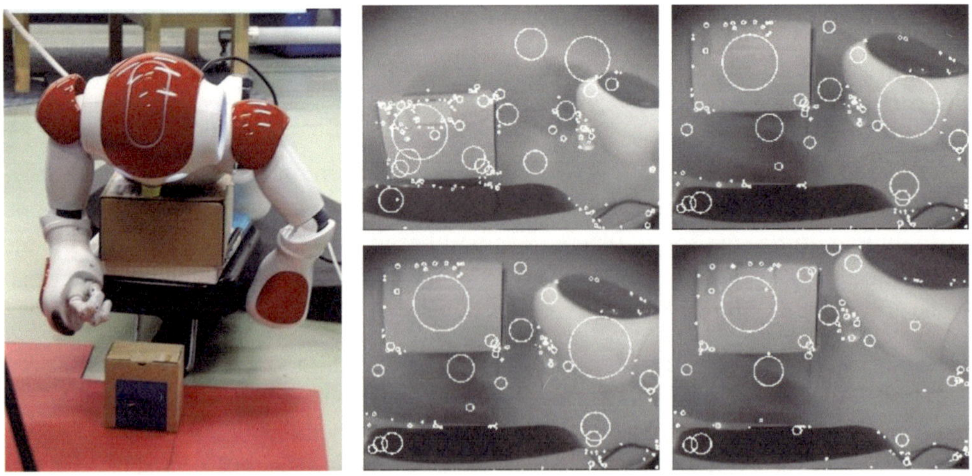

Figure 10. Experimental setup of a humanoid robot with an object (left). Images and SIFT keypoints extracted as circles with different scales indicated by circles (right).

The objective for the robot is to construct a space that provides the following utilities:

1. Estimation of closeness between its hand and the object
2. Prediction of collision between its hand and the object

The first utility allows the robot to plan its motion so that its hand does not to come too close to the object while the robot tries to reach some configuration, avoiding collision with obstacles. The second utility is expected to contribute to the ability to predict collision prior to its motion by integrating it into other techniques, for example, prediction of robot's hand in the image space.

3.2. Manifold learning based on SIFT image features

Manifold learning by LLE is applied to the vectors represented by positions of SIFT keypoints. Each keypoint contains 128-dimensional feature vector that is used to classification and matching to the keypoints in other image frames. By the matching process, a keypoint can be tracked through multiple image frames given that it is extracted in those images. However, in the application of robot motion sequence, each feature vector corresponding to a keypoint is not consistent through sequences of image frames. The arm, which consists of serial links, inevitably changes its posture while it is moving toward a certain configuration. By assuming that each keypoint tracks a certain part of the arm, we proposed a method for matching and labeling using self-organizing map (SOM) [23].

Although feature vectors of a keypoint differ by the change of the robot's configuration in the image frames, it is likely that those feature vectors in images with small differences in image pixel level are similar. By using topological neighbor of SOM generated by image pixel information, correspondence between keypoint labels can be found. By finding correspondence between neighbor nodes, labels that correspond to the same part of the real world are integrated into one label.

3.3. Motion generation based on manifold learning

Dynamic programming with discrete state representation [10] is applied for motion generation. The state for motion generation is defined by the joint angle space. The discrete state is given by discretizing the joint angles of the robot two-dimensional grids. Actions are defined as four directional transitions from a grid to its adjacent grids. Reward is defined as 0 for reaching the desired configuration, −100 for colliding with the object, and −1 at every step otherwise. Collision with the object is predicted using the obtained LLE representation as described below.

3.4. Simulation results

We first tested basic property of LLE in conditions similar to the experimental problem setting. Virtual keypoints were generated as shown in **Figure 11(a)**. As an assumption, an object and the robot hand is captured in an image frame with the size of 400 × 400 [pix]. There were 10 keypoints to be detected on the object, 10 on the robot hand, and 5 in the background. Both the positions of the object and the hand were varied randomly with uniform distribution.

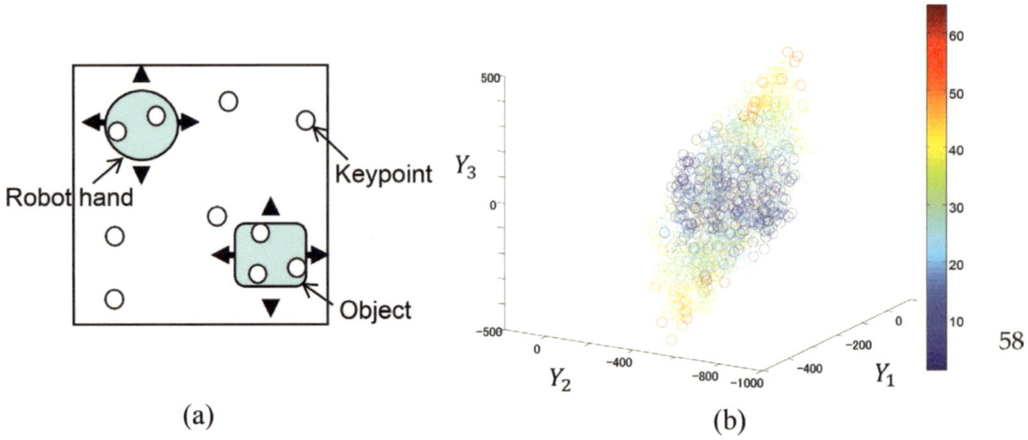

Figure 11. Simulated keypoints (a) and result of LLE with distance information between robot hand and object (b).

Total number of images was 1000. Number of keypoints was 25. To simulate matching error of keypoints, position information of 10% of the keypoints in the data vector was removed.

The result of mapping by LLE is depicted in **Figure 11(b)**. Y_1, Y_2, and Y_3 in the figure correspond to low-dimensional vector y and hence they do not have units. The colors of the points denote distances between the object and the hand in the corresponding images, where the original distance information in pixel with maximum 550 pixel was converted to 64 levels. It can be seen in the figure that one direction in the feature space reflects the distance between the object and the hand.

3.5. Experiment of LLE mapping and motion generation with real images

The three-dimensional mapping constructed by the proposed method is depicted in **Figure 12**. Each point, indicated by a circle or a cross, indicates a vector obtained by converting the image feature vector by LLE. A cross denotes an image corresponding to a situation where the hand contacts with the object. A circle denotes an image without any contact. It can be seen that in the space, crosses are concentrating around a certain region. Distance between the object and the hand, however, could not be clearly seen in the obtained map.

For verification, some test images that are independent from the training process of LLE mapping generation were mapped onto the generated space. Test samples are drawn by boxes in the figure. Corresponding images are also displayed. It is observed that the image with its robot hand, the most distant from the object, is located in the space at the furthest position from the region of the dense crosses. Images with its hand closer to the object are located also closer to the "contact" region. But there is a jump at the last step to contact with the object into the region with dense crosses. Thus, the spatial relation between the hand and the object was reflected to a certain level, but not directly reflecting the distance between the hand and the object in the real world.

Classification of collisions was also evaluated based on the generated map information. Using the mapping collision between the hand and the object was predicted by whether an image is included in the sphere whose center is the average of the samples indicated by the crosses. The optimal radius was set as r = 0.74, which was found empirically so that the discrimination

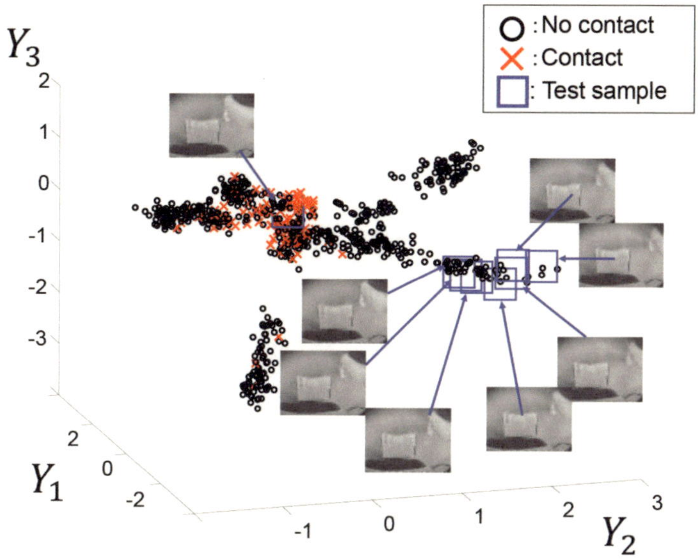

Figure 12. LLE mapping by real images with test samples.

performance was the best. The classification result is shown in **Table 1**. For comparison between linear and nonlinear methods, a linear mapping was also implemented. The classification result based on the mapping principal component analysis (PCA) is depicted in **Table 2**. It can be seen that nonlinear mapping brought conspicuous difference of classification performance.

A sequence of snapshots of motion generated by DP is shown in **Figure 13**. Grid sizes for the discrete state space were set as 8 × 12. Collision was predicted by a correct recognition result for images adopted in **Table 1**. (1) in the figure denotes the initial configuration of the robot hand. The tip of the hand is located above the object in (11), corresponding to the target configuration. It can be seen that the robot hand could reach a destination while avoiding collision with the object, given that an appropriate evaluation of closeness (or collision) to the object is achieved.

	Collision (%)	No collision (%)
Recognized as collision	95/115 (82.6)	111/617 (18.0)
Recognized as no collision	20/115 (17.4)	506/617 (82.0)

Table 1. Prediction of collision with LLE.

	Collision (%)	No collision (%)
Recognized as collision	63/115 (54.8)	132/617 (21.4)
Recognized as no collision	52/115 (45.2)	485/617 (78.6)

Table 2. Prediction of collision with PCA.

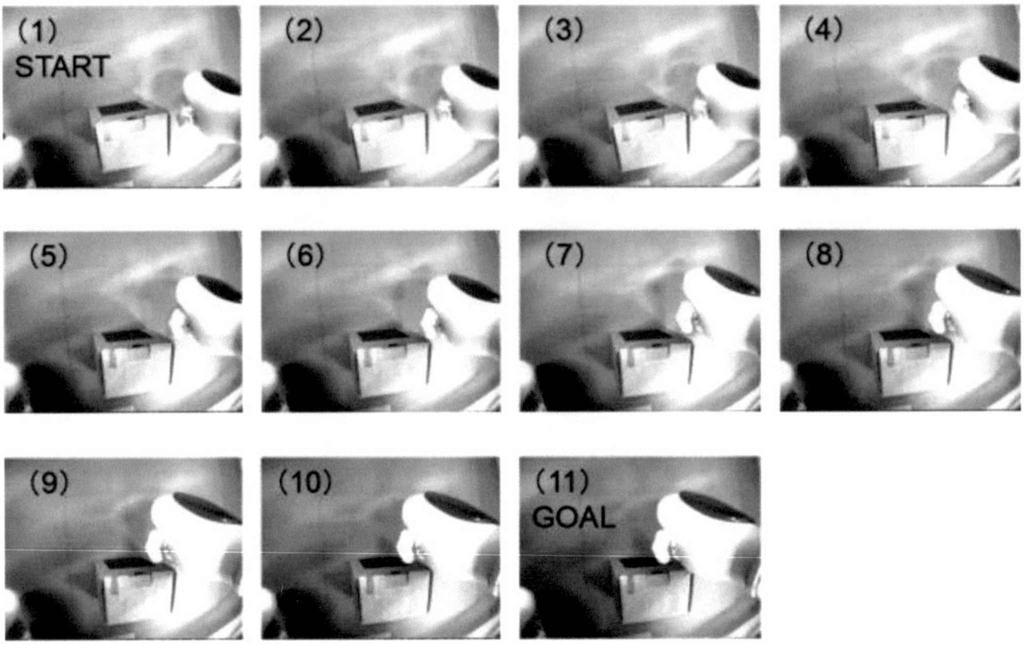

Figure 13. Snapshots of motion a motion sequence achieved by the proposed motion generation.

4. Conclusion

Two kinds of application of manifold were presented in this chapter. In the first application, coordinate systems obtained from sensor signals are directly used for motion control of the robot. In the second application, an intermediate representation, spatial relation between the robot hand and the object, was built using a manifold learning method. One important advantage of these approaches, in comparison with the end-to-end motion learning approaches such as deep learning (e.g., [24]), is that we can analyze and evaluate the obtained representation. In order to apply the approach of manifold learning to more complex robot motion problems, it will be required to consider multiple resolutions, disappearance of features (as discussed in [25]), multiple relations among variables (e.g., discussed in [26]), and connecting different modalities with discontinuous dynamics (such as contact and noncontact switching).

Author details

Yuichi Kobayashi

Address all correspondence to: kobayashi.yuichi@shizuoka.ac.jp

Shizuoka University, Hamamatsu, Japan

References

[1] Asada M, MacDorman K, Ishiguro H, Kuniyoshi Y. Cognitive developmental robotics as a new paradigm for the design of humanoid robots. Robotics and Autonomous Systems. 2001;**37**:185-193

[2] Stoytchev A. Some basic principles of developmental robotics. IEEE Transactions on Autonomous Mental Development. 2009;**1**(2):122-130

[3] Thrun S, Burgard W, Fox D. Probabilistic Robotics. 2005

[4] Fox D, Thrun S, Burgard W, Dellaert F. Particle filters for mobile robot localization. In: Doucet A, de Freitas N, Gordon N, editors. Sequential Monte Carlo Methods in Practice. Springer, New York, NY: Springer Verlag; 2001. pp. 499-516

[5] Elfes A. Using occupancy grids for mobile robot perception and navigation. Computer. 1989;**22**(6):46-57

[6] Saul LK, Roweis ST. Think globally, fit locally: Unsupervised learning of low dimensional manifolds. Journal of Machine Learning Research. 2003;**4**:119-155

[7] Kobayashi Y, Kurita E, Gouko M. Integration of multiple sensor spaces with limited sensing range and redundancy. Inernational Journal of Robotics and Automation. 2013; **28**(1):31-41

[8] Luo ZW, Ito M. Diffusion-based learning theory for organizing visuo-motor coordination. Biological Cybernetics. 1998;**79**:279-289

[9] Luo ZW, Ando H, Hosoe S, Watanabe K, Kato A. Spatial generalization of optimal control for robot manipulators. Journal of Robotics and Mechatronics. 2001;**13**(5):533-539

[10] Sutton R, Barto A. Reinforcement Learning. Cambridge, MA: MIT Press; 1998

[11] Kaelbling LP, Littman ML, Cassandra AR. Planning and acting in partially observable stochastic domains. Artificial Intelligence. 1998;**101**:99-134

[12] Andrew McCallum R. Instance-based utile distinctions for reinforcement learning with hidden state. In: Proceedings of the 12th International Machine Learning Conference. 1995

[13] Wierstra D, Foerster A, Peters J, Schmidhuber J. Solving deep memory POMDPs with recurrent policy gradients. In: Proceedings of the International Conference on Artificial Neural Networks ICANN'07. 2007

[14] Kass M, Witkin A, Terzopoulos D. Snakes: Active contour models. International Journal of Computer Vision. January 1988;**1**(4):321-333

[15] Fukunaga K. Introduction to Statistical Pattern Recognition. San Diego, CA, USA: Academic Press Professional, Inc.; 2013

[16] Schölkopf B, Smola A, Müller K. Nonlinear component analysis as a kernel eigenvalue problem. Neural Computation. 1998;**10**(5):1299-1319

[17] Tenenbaum JB, de Silva V, Langford JC. A global geometric framework for nonlinear dimensionality reduction. Science. 2000;**290**(5500):2319-2323

[18] Prankl J, Zillich M, Vincze M. 3d piecewise planar object model for robotics manipulation. In: Robotics and Automation (ICRA), 2011 IEEE International Conference on. 2011. pp. 1784-1790

[19] Kobayashi Y, Matsui R. Manifold learning approach toward constructing state representation for robot motion generation. Transactions on Computational Collective Intelligence XXIV, LNCS. 2016;**9770**:101-116

[20] Lowe DG. Object recognition from local scale-invariant features. In: Proceedings of IEEE International Conference on Computer Vision. Vol. 2. 1999. pp. 1150-1157

[21] Kobayashi Y, Okamoto T, Onishi M. Generation of obstacle avoidance based on image features and embodiment. International Journal of Robotics and Automation. 2012;**24**(4): 364-376

[22] https://www.softbankrobotics.com/emea/

[23] Kohonen T. Self-Organizing Maps. Tokyo, Japan: Springer Press; 1995

[24] Mnih V et al. Human-level control through deep reinforcement learning. Nature. 2015;**518**:529-533

[25] Somei T, Kobayashi Y, Shimizu A, Kaneko T. Clustering of image features based on contact and occlusion among robot body and objects. In: Proceedings of the 2013 IEEE Workshop on Robot Vision (WoRV2013). 2013. pp. 203-208

[26] Kobayashi Y, Hosoe S. Planning-space shift motion generation: Variable-space motion planning toward flexible extension of body schema. Journal of Intelligent and Robotic Systems. 2011;**62**(3):467-500